Customer-Driven Marketing

Customer-Driven Marketing

Lessons from Entrepreneurial Technology Companies

Edited by

Raymond W. Smilor
IC² Institute and
Graduate School of Business,
The University of Texas at Austin

Lexington Books
*D.C. Heath and Company/Lexington,
Massachusetts/Toronto*

Library of Congress Cataloging-in-Publication Data
Customer-driven marketing : lessons from entrepreneurial technology companies / edited by
  Raymond W. Smilor.
    p. cm.
  Includes index.
  ISBN 0–669–21128–1 (alk. paper)
    1. High technology—Marketing. 2. Technological innovations—
  Marketing. I. Smilor, Raymond W.
HC79.H53C87   1989
620'.0068'8—dc19   89–2519 CIP

Published simultaneously in Canada
Printed in the United States of America International Standard Book Number: 0–669–
21128–1
Library of Congress Catalog Card Number: 89-2519

The paper used in this publication meets the minimum requirements of American National
Standard for Information Sciences—Permanence of Paper for Printed Library Materials, ANSI
Z39.48–1984. ∞™

Year and number of this printing:

89 90 91 92 10 9 8 7 6 5 4 3 2 1

Contents

Preface   vii

Acknowledgments   xi

Foreword: Turning Devices into Products   xiii
William H. Davidow

Part I: Success Factors for High-Technology Marketing   1

1   Why High-Tech Products Fail   3
Regis McKenna

2   Product Newness and Market Advantage: Risk Control through Technological Innovation   15
Pier A. Abetti and *Robert W. Stuart*

3   Marketing Lessons from Silicon Valley for Technology-Based Firms   33
Albert V. Bruno

4   America's Fastest-Growing Company: Compaq's Market Creation Strategy   45
Rod Canion

Part II: Accelerating the Technology Innovation Process   55

5   Breaking the Barriers to Technological Innovation   57
S. Ram and *Jagdish N. Sheth*

6   Marketing to Nonexistent Markets   79
John K. Ryans, Jr. and *William L. Shanklin*

7   The Marketing Challenge: Factors Affecting the Adoption of
High-Technology Innovations   91
Peter J. LaPlaca and *Girish Punj*

8   Fast-Track Marketing: Stages of Growth in Ashton-Tate   109
Edward M. Esber, Jr., and *Michael Stone*

**Part III: Marketing Techniques for Technology
Companies   119**

9   Marketing Technology-Intensive Products to Industrial Firms:
Developing a Service Orientation   121
Rajendra K. Srivastava

10   Public Relations Techniques for High-Technology
Start-Ups   137
Dennis Lewis

11   Determining Whether Telemarketing Can Sell
High-Technology Products and Services   147
Joel Leider

12   Technology Forecasting, New Product Development, and
Market Evolution   159
John H. Vanston and *Donna C. L. Prestwood*

Glossary of Technology Forecasting Terms   173

Index   181

About the Contributors   185

About the Editor   189

Preface

Traditional marketing approaches maintain that one cannot create a market; that marketing professionals can only take advantage of already existing markets. This book argues differently. Successfully marketing technologically innovative products and services requires a market-creation attitude on the part of marketing professionals. Rather than seek only shares of existing markets, high-tech marketers must be customer driven to build relationships, find new users, and develop better and expanded applications, all of which generate new market opportunities. In the process, they must overcome several problems unique to marketing technologically innovative products and services.

Technology/Market Gap

Technical innovations often result in a gap between technology and the market. The gap arises because of two perspectives. On the side of those who produce the technology, there is a great understanding of the technology. This intrinsic understanding leads to high interest in the innovation and a desire to push something new into the marketplace. Potential users or buyers, on the other hand, often lack intrinsic knowledge about the technology, may be uninformed, or may have various levels of sophistication in understanding. Consequently, they have lower interest in trying something new. They are satisfied with what they have been using and are extremely concerned about cost. So how do marketing professionals bridge the gap?

It requires a new way of thinking about customers and marketing and involves greater risk for high-tech ventures. The more technologically innovative the product or service, the less effective traditional approaches to the market tend to be. For example, forecasting demand for technological innovations is difficult because potential buyers may not have a clear frame of reference for the new technology. In addition, for many innovations, applications may be uncertain and encompass a variety of industries.

For traditional products and services, product life cycles may be twenty, thirty, or forty years once they are introduced into the market. With technological innovations, a product may have no more than eighteen months before technical obsolescence, competition, or changing customer needs alter the market. So speed of development poses special problems in going to market.

Finally, there is reluctance to buy early-generation technologies. Potential buyers tend to wait for later generations, feeling the technology will work better and cost less.

The lessons from entrepreneurial technology companies described by leading practitioners and scholars in this collection of papers show how to deal with these challenges. High-technology marketers must be more aggressive and innovative; must rely more on judgment and intuition; and must be more entrepreneurial in developing and expanding market opportunities.

New Imperatives

Professional marketers must involve potential customers as early as possible in the research and development stage. Market research must be qualitative as well as quantitative. Initial marketing efforts must focus on innovators—those individuals or companies more willing to try something new first. Through demonstrations, potential buyers must be vicariously or directly involved in the process of understanding the technology and its applications.

High-technology marketing requires professionals to educate buyers to emphasize the benefits rather than only the attributes of technological innovations. This is because high-technology development is based on the scientific method—on logic, analysis, and reasoned, rational progression. But customers are not!

Consequently, great marketers are students of human moves. They are constantly watching, evaluating, sensing, interacting with, responding to, and anticipating customers.

Educating the Marketplace

But there's an important twist in this observation for those responsible for marketing technologically innovative products and services. These students are also teachers. They not only are learning from the marketplace, but also are constantly educating it. They do more than sell or promote. They enlighten. In fact, the learning is inseparable from the teaching. One constantly reinforces and expands the other.

Education is not an end. It is a process. It requires an ability to communicate appreciation for, knowledge about, and experience in a subject. By conveying benefits and values, the process instills enthusiasm, recognition, and even enjoyment. It engenders a willingness to become involved.

Great high-tech marketers, therefore, constantly are educating and being educated by the marketplace.

When a product solves a real problem or meets a real need, it reflects a deep appreciation for and knowledge about the customer. When the customer is educated to the product's values and benefits, then the result can be the "ah ha" sensation of discovery that comes with customer-driven products.

Sensing the Customer's Pulse

The home-run products, the company-makers, are those that spring from sensing the pulse of the customer and then matching real product differences with real customer needs. The requirement for educating the marketplace thus has broader implications for high-technology marketing. It is inextricably tied to the process of providing good service and customer support. Learning, for the marketer and the customer, continues after the sale. Indeed, training and education are inherent parts of technologically innovative products.

Becoming educated to the marketplace requires quantitative information and qualitative insights. One springs from data, the other from experience. One relies on numbers, the other on judgment. One demands objectivity, the other personal involvement.

Power of Demonstration

Great product campaigns educate largely because they use the power of demonstration. The ability to prove a product, show a benefit, visualize a value, illustrate intangibles, or display differences is the best teaching tool. It's the best not only because it reduces uncertainty, but more importantly, because, if it reflects an understanding of human moves, it also captures the imagination.

Overview of Book

This volume is organized in three parts. Part I identifies a number of factors that correlate with success for technologically innovative products and services. It assesses why high-tech products fail and shows the relationship between product newness and market advantage. It highlights lessons from Silicon Valley firms and presents a case demonstrating market creation strategy.

Part II examines ways to accelerate the technology innovation process. It shows how to overcome market barriers and thus create markets for technical innovations. It assesses key factors affecting the adoption of innovations and presents a case demonstrating the stages of growth in entrepreneurial technology companies.

Part III identifies marketing techniques that reflect a customer-driven strategy. It emphasizes the importance of a service orientation, describes public relations and telemarketing approaches for high-tech companies, and describes technology forecasting methodologies to help move new products into the marketplace.

This book is dedicated to enhancing our knowledge and understanding of marketing technologically innovative products and services. It links theory with practice by weaving the experience of practitioners with the analysis of academics. This approach provides examples for entrepreneurs, insights for marketing professionals, and analysis for scholars.

Acknowledgments

It is a pleasure to thank friends and colleagues who helped make this book possible.

I wish to especially acknowledge two exceptional organizations—the IC² Institute and the Graduate School of Business at the University of Texas at Austin. I have appreciated the support and advice of George Kozmetsky, director of IC², and Dean Robert E. Witt, Associate Dean Robert Sullivan, and Department of Management Chairman Timothy W. Ruefli of the Graduate School of Business.

I appreciate very much the interest, encouragement, and at times editorial assistance of **Mary Kay Allen**, assistant professor, Air Force Institute of Technology; **Francis Bidault**, department head, Graduate School of Business, Lyon, France; **James Botkin**, partner, Technology Resources Group; **Arnold C. Cooper**, professor, Krannert Graduate School of Management, Purdue University; **C. Merle Crawford**, professor, Graduate School of Business Administration, University of Michigan; **Paul Dali**, marketing consultant; **Rohit P. Deshpande**, professor of marketing, Amos Tuck School, Dartmouth College; **Luis V. Dominguez**, chairman and professor, Department of Marketing, University of Miami; **Dan T. Dunn, Jr.**, associate professor of Marketing, Northeastern University; **Peter H. Farquhar**, associate professor of Marketing, Carnegie-Mellon University; **David Ford**, School of Management, University of Bath, England; **David C. Forman**, vice-president of development, Spectrum Training Corporation; **Mark G. Frantz**, president, Frantz Medical Development Limited; **Wesley J. Johnston**, associate professor of marketing, University of Southern California; **Michael Karnow**, editor, Information World, CMP Publications, Inc.; **George H. Kidd**, partner, L. William Teweles; **Thomas J. Kosnik**, assistant professor, Business Administration, Harvard University; **Peter La Placa**, professor of marketing, University of Connecticut; **S. Thomas Moser**, national director, High Technology Practice, Peat Marwick; **Albert L. Page**, associate professor of marketing, University of Illinois at Chicago; **N. Mohan Reddy**, assistant professor of marketing, Case Western Reserve University; **John K. Ryans, Jr.**, professor of marketing and International Business, Kent State University; **William L. Shanklin,** professor of Marketing and International Business, Kent State University; **Nicholas Staveley,** head of Marketing Operations, International Products, British Telecom; **Hiroyuki Torii,** editor, Nikkei High Tech Report, Japan; **Walter K. Weisel,** president and CEO, PRAB Robots, Inc.; **Rolf T. Wigand,** professor of Public Administration and Communication, Arizona State University; and **Peter Zanden,** president, IntelliQuest, Inc.

I am grateful for the support and advice of the late **Wayne Brown,** founder, Utah Innovation Center; **Ian G. Dalton,** director, Research Park, Heriot-Watt University, Edinburgh, Scotland; **Edward M. Esber,** chairman and CEO, Ashton-Tate; **George N. Havens,** chairman, Strategic Research Center; **Pierre Laffitte,** senator of France and president, Technopole Service, Sophia Antipolis, France; **Richard Laster,** president, DNA Plant Technology Corporation; **Modesto Maidique,** president, Florida International University; **Regis McKenna,** chairman, Regis McKenna, Inc.; **Peter Meldrum,** president and CEO, NPI, Inc.; **D. Bruce Merrifield,** assistant secretary for Productivity, Technology, and Innovation, U.S. Department of Commerce; **Everett M. Rogers,** associate dean, Annenberg School of Communications at the University of Southern California; **Mark B. Skaletsky,** president and CEO, Biogen, Inc.; **Eugene Stark,** former chairman, Federal Laboratory Consortium, Los Alamos National Laboratory; and **Milton Stewart,** president, Small Business High Technology Institute.

It has been a pleasure working with Karen Hansen, editor, Susan Geraghty, production editor, and Kelly Brewer, copy editor, at Lexington Books. The manuscript benefited from their valuable assistance.

I thank Elaine Chamberlain of the IC² for her help with the papers. I especially appreciate the work of Linda Teague of the IC² Institute for all her assistance in preparing the final copy of the manuscript.

I am grateful to the authors whose contributions make up this volume for sharing their ideas, experience, and knowledge about high-technology marketing.

My special thanks go to two very promising entrepreneurs, Matthew and Kevin Smilor, for all the lessons they've taught their father, and to Judy Smilor for being a wonderful partner in life.

Foreword
Turning Devices into Products

William H. Davidow

T HE phone rings frequently these days. People involved in high-tech call because they would like me to speak to a group of professionals, a trade association, a conference or maybe even the management of a company. In response to the invitation, my first question is to ask about what? The most frequent request is to talk about marketing—what it is, why it is important, and whether high-tech companies need it.

Since the birth of high-tech, marketing had really been unimportant. Management of fast-track corporations had worried about the factors that made their businesses tick—technology, manufacturing, and quality. If they did these right, the products sold themselves—until recently, that is.

No employee of a high-tech company ever calls a speaker and asks him to talk about why technology, manufacturing, or quality are important or whether companies need these capabilities. These are obvious. What's more, they are tangible. After all, no one can sell a product that has never been engineered and cannot be built—at least not for very long. Customers can precisely measure the quality of products and will not buy them if they do not meet exacting standards.

Why is it so hard to grasp the fact that high-tech companies need marketing as well? One reason is that frequently in the past they did not. They could get away with building a better mousetrap. If they did there were fewer competitors to copy it and confuse the market about it. In this environment, the world would beat a path to their doors.

The second reason is that, while the other disciplines of business are tangible and absolute, marketing is intangible and relative. It has to do with brand image, product positioning, distribution channels, and pricing—the type of issues that can be debated endlessly but never be proved right or wrong. Marketing is the type of subject about which everyone can have an opinion but which no one really understands.

Many companies have existed in the past and did not, by my definition, have marketing. None, however, existed without a manufactured product or service to deliver. Those products or services had to be engineered and produced. Consequently, you have it, marketing has been the appendix of high-tech businesses—something companies could live with or live without.

As best I can determine, many technologically oriented presidents of companies have decided to operate on themselves and excise the troublesome marketing appendix. But in doing so, they have performed a lobotomy instead of an appendectomy. I would like to explain why.

Shifting Battleground

Something is happening today. The old competitive weapons are losing their zing. Technology, manufacturing, and quality are becoming the offensive tools of the past. They are quickly losing ground to the business skills of the future. The tools corporations have traditionally relied on for their success will be inadequate by themselves for future battles. It is important to understand why.

High-tech businesses compete for customers principally through the use of technology, manufacturing, quality, marketing, and service skills. The businesses that win, win because they are either significantly better than a competitor in one or more of these areas or because they are significantly different from a competitor. In other words, by massive differentiation, they offer a product with little or no direct competition.

If, in fact, one looks at the way companies have competed in the past, the pattern has been remarkably similar. First, they competed on the basis of technology. Frequently a company or a country would possess a unique set of skills; it could do what no other competitor could. It could make steel, TV sets, or transistors, and no one else either knew how or had the capability. Companies in this position owned the market; they had no competition.

But, after a while, if the skill were valuable, other competitors learned how to build similar or identical things. The technology diffused around the industry or world. Then the battleground shifted from my technology versus yours—or the absence of your technology—to my cost versus your cost. In case you happen to feel this is a recent trend, I should like to tell a story about the industrial revolution in Britain. I have borrowed this history from Robert Reich's book, *The Next American Frontier:*

> This was a time when inventions bred new inventiveness. Each new process or machine provoked additional insights and inspired further refinements.
>
> The output of many commodities, particularly coal, iron, and textiles, skyrocketed. The productivity of a British spinning worker increased three-hundred-fold between 1750 and 1825.
>
> Cheap British goods flooded Europe and North America, forcing other economies to adjust. Some sought niches in order to survive. Others spe-

cialized in selling Britain raw materials. Many tried to compete with Britain head-on by adopting British inventions and combining them with cheap local labor.

Despite Parliament's efforts to maintain Britain's lead by prohibiting the export of machinery and blueprints and the emigration of skilled artisans, British inventions soon found their way to Holland, France, Germany, America, and beyond. Industrial spies smuggled out trade secrets. Scientific journals published detailed blueprints.

We all know the rest of the story. It is remarkable to me that if I substitute the United States for Britain and talk about the 1960s or if I substitute Japan for Britain and refer to the 1980s, the text applies extremely well. In the year 2010, I might be substituting words like Korea and Taiwan as well.

Key Trends in High-Tech

The spread of technology causes the arena of competition to shift quickly from technology to manufacturing efficiency. Customers become vitally concerned with price, and a supplier's viability depends upon its cost structure. If businesses are to continue, they have to find a way to achieve cost parity with competitors, either by reducing their own cost or through some sort of artificial means such as trade barriers.

Today, most developed countries see high-tech as an extremely important way to create value-added products for export. High-tech has become a key component in their economic strategy and is seen as one tool to increase and maintain their high standard of living. Alarmingly, they are no different in this view than a number of developing countries, such as Korea, Taiwan, India, Brazil, and the People's Republic of China, many of which are burdened with heavy debt and view high-tech in the same way. These low-cost manufacturing regions are, therefore, struggling to add high-tech manufacturing capacity so they can compete effectively.

My message is that it is going to be increasingly difficult to gain either a manufacturing or technology advantage. Observe, for example, what is going on in Japan. Lifetime employment is starting to break down as companies with overcapacity lay off people. Highly vaunted Japanese management skills are being tested in real world competition as Japanese corporations struggle with a realistically valued yen. Not surprisingly, the Japanese are finding they were better managers at 250 yen to the dollar than at a 150 yen exchange rate.

When companies become relatively equal, based on technology and cost, it is easy to imagine the competitive arena shifting again, this time to quality. We have all observed this. We now have German quality, Japanese quality,

Hewlett-Packard quality, American semiconductor quality, and quality as Job 1 at Ford.

I would hasten to add that in high-tech, we are also observing another key trend toward standardization. Customers are demanding interchangeable mechanical components, semiconductor memories, videotapes, computers, operating systems, and switching equipment. Think of the world of high-tech. Area after area within it is increasingly being populated by similar, if not identical, standard products using the same technologies, manufactured at comparable costs and, for the most part, meeting the same exacting quality standards.

Commodity Products

A product becomes a commodity when similar products are available from a number of sources. Unfortunately for many companies, high-tech products are increasingly becoming commodity products sold on the basis of price, quality, delivery, and service. They have rapidly transformed from unique works of art to high-tech wheat, soybeans, cocoa, and pork bellies.

This may appear to be an overly harsh assessment of the situation. After all, many products possess unique features that make them different in the manufacturers' eyes. But what really matters is whether the customer perceives these differences as extremely important, desirable, of little value, or possibly for his or her own reasons, undesirable. Unless the customer perceives a feature as essential, other product alternatives may be available. If the customer has many functional alternatives from which to choose, a unique product really is a quasi commodity to him.

Peruse, for example, the business section of the Sunday newspaper and tour the personal computer clones. One comes with a 10.0 megahertz 286 and another with a 12.0. Someone else has a 22 megabyte disc compared with a 20. Another company uses amber phosphors to compete with the other someone else's green, and one claims six I/O slots to the competitors' five. All have a footprint smaller than Cinderella's glass slipper. But the bottom line, I suspect, is price, quality, delivery, support, and service. There you have it: a $5,000 PC whose computation power would have cost you $5 million twenty years ago sold on the same basis as a bushel of corn to the Russians.

If the opportunities to win using technology, manufacturing expertise, and product quality have become more limited, then successful business must find new weapons to use. Certainly one of the most potent is marketing.

Marketing is thus becoming more important. However, it is not because other areas of expertise are unimportant. The problem is that too many competitors are getting good at all these disciplines and so, to beat them, a company must look to others to gain an advantage.

What Is Marketing?

What is marketing really? I must confess it was only recently that I discovered what it was myself, and I had been in the field for more than twenty years. It was only when I wrote a book on marketing that I discovered that my view of marketing had been too limited. The dictionary says marketing consists of "all functions involved in transferring the title of goods from the seller to the buyer." Now I am sure this definition will help marketing professionals, and by using it, they will be able to provide better direction to marketing employees or do a better job of analyzing marketing functions! When I first read the definition, I thought Webster had gone nuts. In truth, as arcane as the definition is, it does contain a key phrase, that being "all functions."

As I thought about those words, I realized many people, myself included, frequently thought of marketing as involving too few functions. That frequently caused problems. For example, there is a great tendency to feel that sales and distribution are not part of marketing or that marketing involves only high-level strategic work and not mundane and boring functions such as order processing and customer service.

In my career, I have been amazed at how many times I have explained to sales managers what product marketing functions do. I have been appalled by the disdain with which marketing strategists treat salesmen and distributors because they do not perceive them as an important part of the marketing process. I had been surprised to find that many small high-tech companies consider the chief financial officer as the person who does the marketing because he prices the product. I finally came to realize that Webster's definition was better than any I had ever heard.

As I reflected on the words "all functions," I began to ask, "What, then, is a product?" I reached the amazingly obvious conclusion that a product is simply all things a customer purchases. At that point, I was struck with the revelation that a product is the device or service a company produces plus "all functions" involved in making it appealing, available, and useful to the customer.

The problem many high-tech managements have is that they believe that most customers buy what is produced in the factory. As a matter of fact, most customers are so focused on the product or service they assume they buy it as well. In truth, as high-tech has matured, all the functions involved in making a device appealing, available, and useful to the customer are becoming an increasingly important component of the sale.

What does one purchase when he buys a product? Consider a suit, for example. I know where I purchased every one of my suits. My secret is not my good memory, but rather a habit of buying all my suits from the same store. If I had a good memory, I could say who made my suits as well. In all honesty, I don't care. I trust the store to pick a high-quality supplier.

When I consider buying a suit, some factors are especially important in influencing my decision, such as the location of the store (it's close to my home). I like to try on a suit after it has been tailored so I know it fits well. I once bought a suit at Brooks Brothers in San Francisco, which is thirty-five miles from my house. When it arrived it did not fit well, so I had to drive seventy miles round trip to get the problem fixed. Consequently, a store's convenient location is important to me. I also enjoy the friendly relationship I have with the salesman, which I have built up over twenty years.

In reality, I don't buy the suit at all. I purchase the distribution channel and take whatever brand it serves me. In this case, the product is the suit plus the distribution channel.

Device Plus Marketing Equals Product

For purposes of definition, it is important to think of a product as being composed of two parts: the manufactured thing or service, plus all other functions involved in making it appealing, available, and useful to a customer. I would, for the purposes of brevity, like to refer to the manufactured thing or service as a device and the functions involved in making it appealing, available and useful, as marketing. That enables me to shorten the definition a bit. When I do, a product becomes a device plus marketing.

Many executives may feel that I have used a trick example of a suit, a typical consumer product, to oversimplify the problem. They will undoubtedly argue that high-tech is more complex, and so the analogy does not strictly apply. My response to this is that my example is becoming more and more applicable every day, especially as more and more functionally equivalent products become available.

For standardized products—such as semiconductor memories that all meet similar specifications—price, quality, distribution, and service are the key ingredients determining the purchase decision.

The key to IBM's success is its brand image, service, and broad distribution. The price performance of its technology is not a significant factor. What keeps Lotus on top? I suspect today it has more to do with brand recognition and distribution than product superiority. I know the telephone systems or copiers most companies purchase have more to do with who gets there first with an acceptable alternative than who has the best product on the market.

Let me demonstrate this with a few personal experiences from my own career in marketing. My career at Intel was constantly plagued by having the best device early in a product life cycle and then after a year or so having this early lead evaporate like water on hot pavement. Motorola did it twice to me with the 6800 and 68000 and Fairchild undid me once with the 3870. That I refer to these experiences in the first person shows how personally I took these insults.

Motorola, the company with bat wings for a logo, represented a type of occult force lurking behind every data sheet, attempting to ruin my marketing career. My first real experience with the Moriartys of the microprocessor field came when they destroyed the technology lead of the 8080, the world's first high performance 8-bit microprocessor. We were having a field day when they announced a device that was faster, cleaner in architecture, and used but one power supply as opposed to three. What did we do? We FUD-DED them. We sold the company behind the microprocessor—Intel's credibility. Suddenly, Motorola designs were disappearing as Fear - F, Uncertainty - U, and Doubt - D, the basic ingredients in FUD, spread throughout the customer base. Our product became the 8080 plus FUD.

Within four years after their annihilation, they rose again, this time reincarnated in a 16-bit microprocessor with a symmetric architecture, linear address space, and higher performance. Our beloved 8086 quickly fell into disfavor until we mobilized our sales force. While on their missions, they discovered a serious lack in Motorola's support infrastructure. Their customers were having difficulty applying the 68000 effectively. Our product quickly became the 8086 with a strong field sales organization plus application support. The resulting 85 percent market share was a pleasure to behold.

But the final indignity was when Fairchild, a nonentity in the microcomputer business, knocked off the perfect microprocessor, the 8048. We had planned it with every feature a customer ever dreamed of except low price and enough I/O pins. Then, with no adequate warning, the losingest bush league team in the semiconductor business one upped the great Intel with a product at half the cost and better I/O. We watched order after order vaporize and decided to segment the market and find small niches where we could win. We held on selling support and ease of use to a broad base of small applications and finally encircled the competition and won. Our product this time was the 8048 plus a customer-support strategy.

Intel's technologic leads in the microprocessor business never lasted very long. They were always great while they did, but they were always ephemeral. The other product dimensions—things like FUD, strong distribution, and customer support became key to maintaining momentum.

At one time, high-tech customers knew everything going on in their industry. Today the industry has so expanded that they not only do not know, but they cannot know. With the product explosion, information explosion, and the rapid rate of technological advance, a person cannot know everything anymore. He only knows a small fraction of what is going on and, increasingly, that fraction is determined by the ability of suppliers to dominate the communications channel to the customer and block out the messages of other suppliers.

When I graduated with a Ph.D., one could find out everything that was going on in the electronics field by reading *Electronic News, Electronics,* and *Datamation.* The only computer conferences one had to attend were the Spring and Fall Joint Computer Conferences. No one worried about developments in Japan or Europe because these areas were no-tech instead of high-tech. Today, I throw out more information each day than I used to get in a month. The communications channel has become cluttered.

A product is composed of hundreds of factors but the most important are shown in table 2–1. The device is what gets engineered and manufactured. The other five items are added by marketing.

The Rising Cost of Marketing

Increasingly, high-tech businesses are finding that the cost of marketing the product often approaches or exceeds the manufacturing cost of a device. To see this, look at the price the customer pays for a product and figure out what percent of that is expended on the marketing function and what percent is spent on manufacturing and development.

Digital Equipment Corporation spends about 15 percent of its revenue on marketing. About 30 percent of its revenue is derived from product-related services—support, maintenance, education, and so on, all of which are a part of marketing. That's 45 percent of its income.

Or think of a personal computer. Here, 30 percent of the purchase price goes to the retailer and probably another 5 percent to 10 percent is spent internally on marketing functions. In computer-aided engineering businesses, marketing costs run over 25 percent of sales and service revenues are on the order of 15 percent.

For these examples, the cost of marketing about equals or exceeds the manufacturing cost. For items like the lowly semiconductor, internal marketing and sales functions run about 10 percent of the sales price, and external distribution functions can add another 20 percent to 25 percent. So for the ultimate commodity, the marketing costs are about one-half the manufacturing cost. And I think it will cost more to market high-tech products in the future.

Many high-tech companies spend a great deal of time trying to figure out how to reduce the cost of building the device. They spend millions on automation and move production overseas. But they spend very little time worrying about marketing costs. The problem of reducing them is complex; in fact, executives often do not want to do that at all. They may want to grow them. The reason is that certain markets may have high marketing access costs, and if one cuts the marketing costs, he will not be able to access the market.

To find out about this, just ask semiconductor manufacturers who tried to cut their costs by cutting distribution margins. They did cut their costs, but, frequently, they cut themselves out of the market as well.

High-Tech Dichotomy

Businesses are becoming more and more preoccupied with reducing costs. At the same time, the area with which they have the least sympathy—marketing—may have to grow disproportionately. Maybe someday we will look and be astounded to find that it costs 60 percent of the price of some high-tech products to effectively market them. If this happens, we will all remember the days when it cost only thirty cents on the dollar to get the product to the customer.

Marketing is the function that sells the device; it is a very important part of the product the customer buys. If a company does not have marketing, it may not have a product at all.

We in high-tech constantly engineer and manufacture devices that are better than the competitors'. Unfortunately, they do not sell well because they lack the marketing component to turn them into a product.

In a strategic discussion about a company and the device it produces, it is essential to remember that (1) Customers buy products, not devices; and (2) Marketing turns devices into products. The cost of transforming a device into a product can be very large indeed. If executives perceive the ascendancy of marketing in high-tech companies, maybe it won't be so painful to spend the money. What's more, some may even become as enthusiastic about marketing as I am and claim that engineers develop devices, but that marketing invents products.

Part I
Success Factors for High-Technology Marketing

1
Why High-Tech Products Fail

Regis McKenna

I recently read a wonderful book titled *The Fourth Dimension: Toward a Geometry of Higher Reality,* by Rudy Rucker. In the second chapter, Rucker describes a book called *Flatland,* written by Edwin Abbott-Abbott and published in 1884. This is a story of a one-dimensional character —a square—who tries to take a trip into a higher dimension. Of course, Abbott's book is really a satire on nineteenth century society. But it presents some thought-provoking insights into what may well be the state of our technology industries today.

The initial question, says Rucker about *Flatland,* is the problem of how these lines and polygons can see anything at all. If you were to put a number of cardboard shapes on a tablecloth and then lower your eye to the plane of the table, you would really just see a bunch of line segments. How can the flatlanders tell a line from a square? How can they build any idea of a two-dimensional world from their one-dimensional retinal images?

Technology Flatland

In the past few years, we have seen the creation of a technology flatland. Where hundreds and thousands of products are one-dimensional in nature, they have little or no perceptual or intrinsic difference from each other.

There are about 150 IBM PC clones, with some 140 or so suppliers of personal computers in general. There are sixty 5¼-inch disc manufacturers. There are more than four hundred word processing packages and 350 spreadsheet packages in the marketplace. There are an estimated sixteen thousand software producers. And, in the past five years, some five thousand software companies produced twenty-seven thousand different software packages.

U.S. organizations spend between $5 billion and $10 billion on the development of new-technology products that fail. That figure does not include the cost of marketing, sales, promotion, and time lost. Although there is no one decisive cause of product failure, certainly on overriding factor is not being market driven; the differentiation of the product is only in the mind of the developer.

In technology flatland, differences in shape are perceived only by the inhabitants; to the outside world, they remain one-dimensional. Likewise, new-product failure is not the result of innovation, but of the way a company conducts its business and projects itself.

Rucker, in *The Fourth Dimension*, tries to get us to think of dimension as any possible variation or distinction. He suggests there is no reason to limit dimensions of the world to space and time. Part and parcel of every object you see is what that object reminds you of, how you feel about it, what you know about its past, and so on.

"If we make an honest effort to describe the world as we actually live it," says Rucker, "then the world grows endlessly more complicated than any three-dimensional model." There is a feeling that the more we delve into reality, the more we find. Far from being limited, the world is inexhaustibly rich.

The Customer Dimension

I think this applies wonderfully well to the world of technology. New-technology products need to seek a new dimension—a dimension that allows customers of technology to see, feel, enjoy, and become excited about the application of technology. That application will not come about by promotional techniques. It will come about by the development of imaginative new products and the meaningful integration of those products into the workplace.

Ideas and opportunities in the past were certainly many and varied, and there is no reason to believe that we have exhausted mankind's imagination. We understand that technology is a learning experience and that we have to apply the lessons we have learned in a timely manner. Education is of no value unless it is applied.

I first became interested in the subject of new-product failure when I noticed that the companies that I worked with, in particular those in the entrepreneurial ranks, developed second products that more often than not were failures.

If you think about it, most products following enormously successful products run into difficulties. Why? I have read the research being done on product failures; in most of it, the researchers ask the entrepreneurs to fill out a questionnaire that explains why their products fail. It is kind of like asking you why your children are ugly. The conclusions they draw, I think, are misleading. Most of what I am going to write about are my own opinions and anecdotal kinds of information.

But first, let's look at some product successes—for example, the Apple II, the IBM PC, DEC VAX, or Lotus 1-2-3.

Factors in Products' Success

1. These products created new markets or expanded existing markets. They dropped new users into the realm of the technology; they did not simply share existing users.

2. They were not inventions; they were incremental improvements. They were changes in the technology—enhancements of it. They were not inventions like the light bulb or the first automobile.

3. These new products were interdependent with the development of other technologies. For example, though I don't think many people recognize it, the real success of the Apple II was based on its being the first small computer to take advantage of the first miniature floppy disk drives. That, along with the use of the 6502, a relatively new microprocessor, really made the product accessible and usable to software people. Of course, Lotus 1-2-3 was successful because it took advantage of the new memory capacity of the IBM PC.

4. Timing is all important. Most ideas for new successful products are not particularly original. The 8086 was modeled on the 8080. The IBM PC was modeled on the Apple II. The Apple II was modeled on a number of products that were available in the mid-1970s, and Lotus 1-2-3 had VisiCalc to examine. Macintosh was patterned after Xerox Star. Thus, the ideas for such innovations are often widely held, but are just not acted upon. The business that focuses its efforts on product market development before anyone else tends to win.

5. Almost all successful technology products are adaptable. For instance, the Apple II is used in pig farming, in education, and in business. It is used in a wide range of different industries, because it is so adaptable.

Almost all technology products go through cycles of incremental improvements, changes, and adaptations. They must be adapted, altered, enhanced, incrementally improved, increased in value, integrated, serviced, and supported. The whole-value aspects or the value-added aspects of technology products are what make them successful. However, there are many other factors that made these products successful, including luck and good management.

Factors in Product Failure

Let's look at some of the things that make products fail:

1. The product does not create or expand the market. In other words, we have seen incremental improvements that do little or nothing to enhance a product's value to the customer. I think some of the new personal computers are a good example of this; products that offer incremental improvements—perhaps a larger screen or added color or speed.

Unfortunately, these improvements don't make a great enough impact for customers to say, "Now we are all going to use a personal computer." This is not just true of personal computers, but *all* technology products. If the consumer perceives only incremental improvement in the performance, there will be no expansion of that product to new markets.

2. Indecision by management as reflected in product design. Management cannot decide whether to improve a product incrementally, displace it, link two products together, go up or down the spectrum, or build something radically new. We see that indecision reflected in the product itself.

Take, for example, the IBM PC Jr.'s compatibility with the PC. IBM was afraid to make it compatible because it didn't want to displace the first product. It felt forced to make the PC Jr. somewhat compatible and promote it that way, so that people could buy the low end and move into the higher end. The Apple III was similar. It was not compatible with the Apple II, and was, in fact, an attempt to go beyond the Apple II. If it had been successful, it would have displaced the Apple II.

3. The people who developed the first product are no longer in a position to decide and judge the technology and market posture. The entrepreneur and the inventor often take on new roles within an organization because their first product was so successful. They become managers with people responsibilities they didn't have before. They no longer have the time to gather information in the marketplace. Most of the early entrepreneurs have an intuitive grasp for what has to be done. Steve Jobs and Steve Wozniak were hobbyists; they built the PC for themselves. The market and technology were integrated. And that integration disintegrated as the company grew.

4. Arrogance, the initial success of a product blinds management. They develop the attitude that no matter what they do, the market will respond favorably.

5. The idea for the second product occurs in a totally different environment from the idea for the first product. The decision-making process is different. Often, the entrepreneur developed the original idea while working for another company, in a different environment.

In addition, although the competitive environment has changed radically from that in which the first product was developed, most entrepreneurs isolate themselves from the marketplace and make judgments from their new back room.

6. Democracy decides the product. I've seen this a great deal with start-up companies. They give stock to all the key managers in the company and hire people who have never made a product decision in their lives or started a company. Yet these people make decisions as to what the next product will be. Marketing managers, sales managers, and everyone else decides that everyone must have a say and "teamwork" is going to select the next product. This is at the heart of almost all second-product failures that I have seen.

7. The company loses track of the market and takes on promotional or Madison Avenue techniques to replace fundamental marketing. Because these techniques are used by the big companies, they mistakenly are identified as a way to be successful. Most entrepreneurs who start companies have a very close identification with their marketplace and the technology; they have one foot in each. Gradually, however, they withdraw that foot from the marketplace and have perspective only inside their own business.

Reading the Audience

Marketing is becoming more important than ever before in the success of new products, because it is a process that learns from the marketplace. It teaches the company how to adapt and when and where to respond to the needs of that market.

I recently heard a television interview with George Burns, the comedian, in which the interviewer asked George, "What made you and Gracie so successful?" George replied, "Well, our audience made us successful." The interviewer said, "Oh, well, I appreciate your modesty, but you two had some inherent talent." And George said, "No. We had a number of routines in our early days. We used to experiment with them, but there were basically two routines: one in which Gracie was very, very sarcastic and one in which she played the dumb scatterbrain. The audience did not respond to the sarcasm. In fact, they used to boo us, so we altered our routine. We kept developing the scatterbrain routine and our product became the one that we saw the audience responding to." He is a good marketer.

Talent in our business today is a given. You have to have talent to be successful. But the willingness and the ability to read your audience is what marketing and new products are all about. The problems facing technology companies today are not those of technology; they are problems of limited dimensions.

Being Innovative

There are two fundamental kinds of innovation. There is the original: the first Polaroid, the first automobile, the first personal computer, perhaps the first retail computer store. We don't see many of those in our lives; they come along maybe once in a decade.

Most of the innovation that we see is incremental improvement and change, which are too easy to make in technology today. It is only when that incremental change really brings new audiences to use technology that it is of value.

Incremental change always used to do that, but not in today's technology. The automatic transmission in the automobile was an incremental improvement that extended driving to many, many, more people. Too many technology innovations today, however, are simply feature improvements. They do not extend the technology to new audiences and they do not create new markets. Small companies built on such innovations will find they are open to market-share battles with large resource-intensive competitors. Consequently, it is critical not only to evaluate incremental change but also to understand its ability to do things to the marketplace.

Adapting to the Market

Technology products are never born perfect. The success of any technology product depends upon its ability to adapt and change according to market requirements.

For example, computers learn, grow, and change based upon response to a particular environment. All successful technology products go through a series of incremental improvements throughout their product life cycles.

The Apple II success was directly related to its flexibility. The addition of capabilities by third-party software people and hardware vendors adapted the product to many different markets. Incremental improvements were also the basis for performance enhancements that reduced manufacturing costs.

The IBM PC adapted, changed, and likewise found success. All successful minis and mainframes are adaptable to different markets by third-party software people, vertical market specialists, systems integrators, and a wide range of peripheral and software vendors.

This adaptability is what makes a product market driven. We don't simply research the marketplace, find a need, and fill it. Technology has a life of its own. It bubbles to the surface. It is only after it reaches the surface when products and technologies begin to adapt to various market segments that success becomes critical. To be successful, companies have to focus not only on the introduction of a new product, but also on its adaptation process. This process is one of packaging the total solution for a given group of customers or a given marketplace.

Market-Driven Products

Technology innovates, but markets refine. The adaptation of technology and products in response to customer needs is what is meant by being market driven.

Market-driven products are much different from other products in that they are more like a service than a product. A computer, microprocessor, computer network, laser, modem, or other technology-based complex product is at the same time specific and generic. Such products are adapted to many different markets through product changes, software support, and service organizations.

For every engineer designing a technology product, ten people are providing its adaptation process. Such products require a high degree of application knowledge and personalization. Market-driven products rely on word of mouth, reference structure, infrastructure development, adaptability, cost effectiveness, and the credibility and reliability of the supplier.

Complex technical products require an even greater degree of qualitative support: no one buys a system or a technical product without references. In fact, all technology products are part of a reference system.

Many times, the main problem of the entrepreneurial company is that it avoids that initial reference system. Especially in the second-product phase, it no longer relies on the reference system for being successful. I was on a flight a few years ago from New York to the West Coast, and I met a fellow who is general manager of a large industrial company. He told me he required all his managers to use the IBM PC. When I asked him what made him think that IBM products were any more networked than any other computer, he said, "I believe that they are." I asked him if he knew what belief was. It's religion, not business!

Customers Buy Perception

One of the problems we have with new-technology products and new businesses is that they don't have religion, and they must join the religion to be successful. When consumers buy a product, they also buy a perception; they buy an intangible as part of every product. Remember, IBM did not make one single component in the PC—not one. They bought it all from other companies. But what they did was put their perception on it. The more qualitative enhancements or "religion" one can ascribe to a product—that is, high quality, good service, good support, leading technology, use by reputable companies, and success—the more likely it is the product will be accepted.

Qualitative attributes are derived from experiences with the product or the organization launching the product. Quality is communicated only through direct experience. You cannot effectively talk about someone else's quality.

The factors most valuable in establishing a qualitative position are service, product performance, and the integration of the technology into the customer's plans and objectives. The future direction of the product and the economic consequences of the user's decision to use that product are also important.

So, the picture of the future of a company, the future of a product line, and the future of a technology is actually part of the perception that your customer has of you and your product today.

For established companies with a history of success, such as IBM, the mere perception of a complete solution will suffice for a period of time. For new companies and companies entering a new market, if the perception and the reality of the solution are not very close, the resulting positive word of mouth is lost. If customers have good experience with a product, they each tell three other people; if they have a bad experience with a product, they tell ten other people.

Monitoring the Market

The old adage that pioneers get arrows in their backs is indeed true in technology businesses. RCA in television, Fairchild in integrated circuits, Atari in consumer electronics, and Xerox in personal computers, were all once innovators. The ever-present danger to any business is *time to market*.

Innovations create new markets and change existing ones. The only relatively secure barriers to competition are established relationships based on a history of reliable responses. Therefore, new companies or companies entering new markets must respond rapidly to the market response itself; they must assess and adapt to the wants of the marketplace.

Companies must remain close to the market after product introduction, monitoring the likes and dislikes, assessing the wants and weaknesses of their product. This monitoring must extend beyond the pure performance of the product to the total solution—technology, future direction, support, service, extensions of the new generation, and so forth.

Young companies tend to focus on the product rather than on the solution to the system. Further, such companies are hampered by their lack of experience and resources to develop parallel pieces to the system while at the same time providing marketing, sales support, application engineering, and service.

In *Inside the Black Box: Technology and the Economy,* Nathan Rosenberg writes, "The growing productivity of industrial economies is the complex outcome of large numbers of interlocking mutually reinforcing technologies, the individual components of which have a very limited economic consequence by themselves."

Companies must be honest with themselves about the whole product. They must ask themselves, "What is the customer's perception of the whole product?" They must decide which pieces of the solution are strategically important to maintain control over and which pieces they can develop with others as part of the system.

Therefore, companies must begin early to evaluate and plan marketing development. They also must be structured to respond quickly to changing environments. Established companies that are sufficiently well positioned to use their resources to enter markets—and enter markets late—must monitor the reactions of the marketplace to new product entries. That monitoring must be done at close range and extend over a sizable period of time.

Most large companies I have dealt with have, in effect, underestimated and ignored the impact smaller companies have upon their marketplace. They arrogantly assume small companies never will be big enough to challenge them.

Monitoring the marketplace very closely is essential for both big and small companies. Emulating and being driven by your competitors is a good thing—not a bad thing. This is essentially what IBM does when it enters any business. This is what Motorola did when it entered the microprocessor business. This is what the Japanese do when they enter *any* business.

Markets are not monolithic, although most companies, big and small, think that success comes from the mass market. Instead of mass marketing, mass manufacturing companies should develop a market-share mentality. Markets are living, changing, dynamic realities. Old ones die; new ones are born every day. Markets are best looked at from a particular set or knowledge base. All industrial technology markets are vertical and all vertical markets are a knowledge base.

The Niche Strategy

Vertical markets are usually identified as niches in which small companies can find safe havens from their larger, more resource-rich competitors. This concept is very popular. But the concept of vertical markets as narrowly defined, slower in growth, and requiring specific kinds of customized products is really wrong.

Donald Clifford and Richard Cavanaugh in *The Winning Performance* studied some 6,117 mid-sized growth companies in America. These companies had growth rates from late 1979 through the early 1980s of four times the growth rate of a comparative quartile of the *Fortune 500* over a five-year period. Of the companies studied, 74 percent got their start with innovative products, services, or ways of doing business. Their data clearly show that participation in niche marketplaces is more likely to be profitable than participation in larger, more established markets. The authors point out that, for the mid-size company, the niche strategy is not born of necessity, but is instead cultivated by design.

Small and mid-size companies simply do not have the resources or the staying power to fight head-to-head battles on all fronts with large, entrenched competitors. Instead, they seek out niches that are either unknown to the larger potential competitors or too small to attract them.

Through sheer perseverance and dedication to serving their customers better than anyone else—which means knowing their customers' needs better than anyone else—the mid-size companies can capture these niches and protect them even from the largest and most formidable of competitors.

Marketing is an investment. The return on that investment is achieved not by sales alone, but by the ability of the supplier to control the environment of the marketplace. The primary asset of ownership in a marketplace is market understanding and knowledge. When you know the market better than your competitor, you are better equipped to judge strategic product direction based on customers' future needs and solve customers' problems.

The key reason you must stay close to the fast-changing technology marketplace is because judgment is based on intuition. You must have one foot in the market and one foot in technology. And you must understand the limits and potentials of each.

From there, apply intuitive judgment. Now by intuitive, I don't mean that you make a judgment based upon some flash of genius in the middle of the night. Intuition is judgment based on experience. Without that type of experience in today's marketplace, without good management, without sound management experience in technology, more likely than not judgments are going to be erroneous.

Making Successful Products

So, what makes successful products?

1. It is important that corporate and product strategies be aligned. Many companies—particularly between their first and second products—move off center from their corporate direction. They move into areas about which they are not knowledgeable. They focus on competing with themselves rather than competing with the marketplace.

2. Team selection is critical. The people who design the products, including those who provide the marketing input, are crucial in all businesses, big and small. 3M is a major corporation that has been successful in many areas based upon the selection of a team that brings a new product to market. The selection of the team itself, rather than any process or formula, is most important.

3. It's vital to eliminate risks up front. Management has to immediately identify where the risks lie. Many times the risk is not in technology but in marketing. We see more and more technology products that really are fine, usable ideas that have encountered a problem—like the lack of a distribution system, the establishment of which is the major risk. Eliminating the risks inherent in a product—whether they involve patents, technology, marketing, or sales—requires putting them up front, working on them, and investing in them early.

4. Today, a technology company cannot take any competitor or situation in the marketplace for granted. You learn from your competition, and reach a level of performance based on the competitiveness of the marketplace.

Ford automobile is the *second* largest selling automobile in Europe, and it is also the *third* largest selling automobile in Europe! The reason for this is that it has market access. Ford had to compete with automobiles in Europe, with the steering wheel on the *left* side in some countries and on the *right* in other countries. So, the company built products to compete with the local environment to match the competitive environment.

Conversely, the problem with the semiconductor industry is that its lack of market access in Japan did not allow it to compete with the quality levels of Japanese products as they were being developed. I think we have to keep constant watch on competitors in the marketplace to bring our products up to and beyond the competitive level.

5. Focus on differentiation. More and more, we are finding difficulty differentiating technology products. For example, there are more ice cream companies; there are more cookie companies; there are more automobile companies; there are more computer companies; there are more semiconductor companies. For this reason, differentiation is becoming very difficult. We have to discover our point of differentiation and then invest in that. No one can be all things to all people.

6. Remember that the marketplace positions your product. The whole product includes perceptions of your business, because most of today's technology companies, in particular new ones, operate in a fishbowl.

For example, the company's financial performance is part of the perception of your product. How many of us would buy an Osborne product? How many of us would buy a Victor product? It is clear that we judge and buy products based upon the financial performance of a company. In doing so, we judge whether it will have a technological future.

7. Companies must build relationships with other companies—largely because they cannot do everything themselves. IBM, a $50 billion corporation, spends $4.5 billion on research and has about $5 billion cash in the bank. Yet it is establishing more relationships in the marketplace to buy technology and products.

Even IBM has publicly stated that the company cannot do it itself. And yet we find more and more small entrepreneurial companies doing vertical integration and assuming they can build products and not build the close relationships necessary to get to the market quickly.

More than anything else, our inability to accept change and our habit of doing something new without touching customers in a meaningful way inhibits the growth of technology businesses.

References

Abbott, Edwin A., *Flatland,* Pasadena, California: Grant Dahlstrom (1884).

Clifford, Donald K. Jr., and Cavanaugh, Richard E., *The Winning Performance: How America's High Growth Midsize Companies Succeed,* New York, New York: Bantam Books (1985).

McKenna, Regis, *The Regis Touch: Million-Dollar Advice from America's Top Marketing Consultant,* Reading, Massachusetts: Addison-Wesley (1985).

Rosenberg, Nathan, *Inside the Black Box: Technology & the Economy,* New York, New York: Cambridge University Press (1983).

Rucker, Rudy, *The Fourth Dimension: Toward a Geometry of Higher Reality,* Boston, Massachusetts: Houghton Mifflin Company (1984).

2
Product Newness and Market Advantage: Risk Control through Technological Innovation

Pier A. Abetti
Robert W. Stuart

New-product development and marketing are major driving forces of technologically intensive industries—and inherently risky processes. But by understanding the three dimensions of product newness and their relationship to risk, marketing professionals can better use technological innovation to ensure the technical, commercial, and financial success of their new products. (For brevity, the word "product" is used to also include "process," "service," and "system" offerings.) First, our definitions.

Product Newness

Product newness is a relative concept, depending on the knowledge (market, technology, financial, etc.) and attitudes (psychological, managerial, social, etc.) of the evaluator. It may be evaluated differently by the supplier (who may offer his or her "new" version of a product already available on the marketplace; for instance, another pocket calculator with photocells) and by the potential user (who may perceive "new" utility from a mature product; for instance, dental hygiene using baking soda).

This study takes first the viewpoint of the *supplier,* and then explores newness from the viewpoint of sophisticated *users* (or potential users) in the marketplace.

Product newness can be measured along three orthogonal dimensions: market, functions, and technology (see figure 2–1). While market newness and technological newness have been recognized,[1] functional newness has not received much attention in the literature.[2]

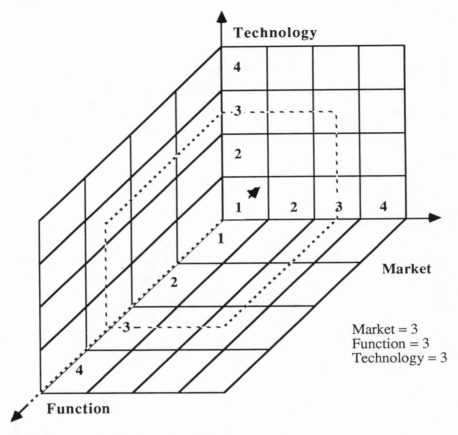

Figure 2–1. New Product with Newness Levels

Market Newness

The firm can expand its scope by offering, with minimal functional or technical changes, its existing products to new, that is, previously unserved, markets. The degree of "market newness" ranges from no change in customers, selling methods, and channels of distribution to a need for acquiring new customers, new selling methods, and new channels of distribution.

Adams has proposed the four-level scale of increasing "Market Newness" shown in table 2–1,[3] to which we have added the example of IBM's potential choices in marketing personal computers.

Table 2–1
Market Newness

Level	Customers	Distribution	Example (IBM Personal Computer)		
1	Same	Same	To Business	-	Direct Sales
2	Same	New	To Business	-	Through Distributors
3	New	Same	To Consumers	-	Direct Sales
4	New	New	To Consumers	-	Through Distributors

Functional Newness

A sophisticated customer purchases new high-technology products because these provide improved or new functional utilities. This customer should be concerned only with the functions performed and their benefit-to-cost ratio, rather than with the underlying technology ("what" the product does, not "how" it does it). Thus, the buyer's role is to specify the desired functions, and the supplier's role is to select the most appropriate technology to efficiently perform these functions.

The firm can expand its scope by offering new functions to the marketplace, without any changes in the market served and the technology used. The degree of "functional newness" ranges from no functional changes to entirely different functions previously not available in the marketplace.

Examples are the first electronic watch, which added the calculating function to the time-keeping and alarm functions, and in-car navigational systems using satellites. Table 2–2 shows a four-level scale of increasing "functional newness," with the example of General Electric's potential choices in developing improved or new medical diagnostic imaging systems.

Technological Newness

The firm should select the most appropriate technology to obtain the desired functional specifications. Established firms have assembled a body of knowledge in the core and supporting technologies[4] embodied in their product offerings. This body of knowledge has been acquired through internal Research and Development (R&D), technology transfer from users or suppliers, or from acquisition of external patents and know-how.

Without changing the functional specifications of the product and the served market, a firm may decide to change the technology embodied in the

Table 2–2
Functional Newness

Level	Description	Example (GE Medical Diagnostic Systems)
1	Incremental functional change of (or addition to) existing company product	X-ray with improved resolution
2	Major functional change in existing company product	Realtime X-ray -- provides pictures more quickly
3	Copy of adaptation of another company's product functions	CAT-scanner -- provides enhanced detail especially in obstructed parts of the body
4	Product with completely new functions	Magnetic Resonance Imaging (MRI) -- provides improved resolution in soft tissue. Also provides biochemical information

product to lower production costs, to increase reliability or maintainability, or to improve performance or efficiency.

The customer sees no change in the functions performed, but may (or may not) profit in terms of reduced prices and higher quality of service.

Table 2–3 shows a four-level scale of "technological newness," with a hypothetical example from the computer memory industry.

Composite Newness

A truly new product will generally present different levels of newness in the three principal dimensions and, on our composite scale, may occupy any one of the $4 \times 4 \times 4 = 64$ possible cells shown in figure 2–1.

Increasing newness in all three dimensions is easily visualized by following the diagonal from the origin of the cube (existing products) to the opposite vertex. This vertex corresponds to an entirely new product, which

Table 2–3
Technological Newness

Level	Description	Example (Computer Memories)
1	Routine engineering changes	Changes in packaging materials and components for cost reduction
2	Application of the company's "state-of-the-art" knowledge	Design of a 256-kilobyte RAM (random access memory) to replace four 64 kilobyte units
3	Development or transfer of a technology previously not available in the company, but available elsewhere	Change of the memory product line from MOS to CMOS technology
4	Development of a new technology not previously used elsewhere	Replacement of the semiconductor-based memory product line by one based on Josephson superconducting junctions

uses as-yet-unproven technology to perform functions not yet available in the marketplace and is offered to customers new to the firm, using new distribution channels.

Success and Risk of Failure

By definition, newness involves change and change involves risk. Risk exists because of uncertainties in predicting technical performance, acceptance by the marketplace, and especially financial returns of a new product.

Today it is generally believed that new products, whether developed by an established firm or by new ventures, suffer significant risk of failure. However, there is substantial controversy in the literature concerning the numerical ratio of successes versus failures.

On the one hand, the pioneer study by Booz-Allen and Hamilton[5] reported that of all product development projects in six major industries (chemical, customer packaged goods, electrical machinery, nonelectrical machinery, metal fabricators, raw material processors), only 14.5 percent, that is, one in seven, lead to commercially successful products.

A recent paper by Mansfield indicates success ratios of 12 percent to 20 percent for products of firms in the chemical, drug, petroleum, and electronics industries.[6]

On the other hand, Crawford, after extensive review of the literature on new consumer product introduction concludes "the best estimate from available sources is that around 35 percent of new products fail."[7] His comparable figure for industrial goods is 28 percent.

Cooper performed recently an in-depth study of 103 Canadian industrial firms and obtained the mean success rate for developed products of 59 percent.[8]

When Does a New Product Get Started?

A major difference between the statistics cited is that Booz-Allen and Hamilton and Mansfield measured new product development projects from their initiation, whereas Crawford and Cooper were examining products that had passed through the development phase and were being introduced to the market.

The new product development process, from idea generation to screening to research and development to production to market introduction has been discussed extensively.[9] It can be compared to an obstacle race, where some runners fall by the wayside at each checkpoint. Thus, the ratio of "successful" products depends not only on the numerator (products successfully introduced in the market) but also on the denominator (number of starting products).

Most high-technology products require significant expenditures in R&D, production facilities, and market research before they are launched in the market.

Thus, a proper starting base would be that portfolio of products for which the company's management has authorized R&D and market research efforts, either formally through acceptance of a new product proposal, or informally through permitting, knowingly or unknowingly, the "bootlegging" of new products.[10]

If this starting point is used, the new product success rates of Crawford and Cooper will be closer to the figures of Booz-Allen and Hamilton and Mansfield, and therefore, the probability of success will be substantially lower than 50 percent.

Another question arises in the definition of "success." Success, like "product newness" is largely in the eye of the beholder, in this case, of the manager who classifies a new product as "successful" or not. From a subjective viewpoint, success can be defined as meeting expectations.

We believe, however, that new product success should be measured objectively in terms of the financial returns returned to the firm by the product over its life. These financial returns should be commensurate with the risk taken in relation to alternative investment opportunities.[11]

Objective and Subjective Risk

The concept of risk is closely connected with that of probability and therefore may be measured objectively (from field data) and subjectively (from the perceptions of the risk-taker). The objective risk of new product failure is complementary to the probability of success, and thus is substantially higher than 50 percent.

The subjective risk, as perceived by managers responsible for new product development, is also significantly high. Little interviewed managers of 152 industrial manufacturing firms in Ontario and Quebec.[12] These managers were asked to rate the "degree of risk to your company in developing this particular new product." In spite of the fact that, according to the managers' evaluation, 146 out of 152 products had been successfully introduced to market, the *ex-post* risk rating was: low 39 percent, moderate 26 percent, and high 34 percent. Thus, even products which had succeeded entailed considerable risk-taking by the firm.

Available quantitative data demonstrate that new product development entails substantial risk. But what is the relationship of risk to product newness?

Product Newness and Risk

Intuitively, the level of risk should be positively related to the degree of product newness; that is, the "newer" the product, the higher the risk of failure.

To our knowledge, the most complete analysis of the relationships between new product success, market newness, and product newness was performed recently in England. Adams studied 63 small manufacturing firms that had introduced 309 products over about six years.[13] Adams asked the question, "Did the new product behave as expected?"

He measured the levels of market newness, defined according to table 2–1, and of product "newness." In this case, product newness is primarily technical newness as in table 2–3, since these small manufacturing firms usually developed their new products or processes to the *customers'* (not their own) functional specifications. Adams' results are summarized in table 2–4.

Adams' data show that:

1. Market newness is more risky than product (technical) newness. The risk of failure for new products sold to the same customers through the same channels was 52 percent. This risk increases with the level of market newness, to reach 92 percent for new customers approached through new channels. This conclusion agrees with Cooper's studies, which emphasize market risk over technical risk and the importance of market research for successful new product introduction.[14]

2. Risk does not show monotonic corrrelation with the level of product newness. The risk is 48 percent for minor technical modifications in existing products and increases to 84 percent for products that are "new" for the firm but are copies of other products already available in the marketplace. However, completely new products show lower risk (68 percent). We will discuss this discrepancy in the next section.

It can be determined then that new product development is a risky business, since about two-thirds of the new products did not meet expectations and risk is *positively* related to newness.

Product Newness and Technological Innovation

We define a technological innovation as a new application of science and technology to a specific market or industry. According to this definition, the application need not be new in an absolute sense, that is, the first ever in the world, as an invention. Rather, the application could be the use of a previously developed technology to new products and/or new markets.

Studies of the factors that affect the success of technological innovations and new industrial products have attempted to establish relationships between the "degree of innovation uniqueness" and the probability of success

Table 2–4
Percent of New Products Not Behaving as Expected

	Level 1	Level 2	Level 3	Level 4	Average
Market Newness	52%	76%	74%	92%	74%
Product Newness	48%	65%	84%	68%	66%

(and thus the risk of failure). Uniqueness in this case is viewed from the market or users' viewpoint as contrasted to "technical newness," which is viewed from the suppliers' perspective.

On the basis of our previous work,[15] we will use a five-degree scale to measure this characteristic, as shown in table 2–5.

According to several reliable studies, based on extensive field research, there is a *positive* correlation between new product success and the uniqueness of the innovation.[16] This means that a truly innovative product has higher probability of success than a "me-too" product.

We are now in a position to explain the lack of monotonic correlation between levels of product newness and risk, as shown in table 2–4. A "level 3" product presents a relatively high degree of newness (from the firm's viewpoint) but a low degree of innovation (from the market's viewpoint). Thus, this product suffers from two simultaneous risks of failure because the firm may lack the appropriate technology, and it is nothing but a "me-too" product in the marketplace.

In contrast, a "level 4" product is new to the firm, but also innovative in the marketplace. The overall risk of failure is lower for the second product, as indicated by Adams' field data.

Obviously, the technological innovation must produce significant value-

Table 2–5
Innovation Uniqueness

Degree	Description
1	Incremental improvement over existing products application of current technology, standardized product, no patent protection, no R&D.
2	Significant extension of product characteristics with original adaptations of available technology, products with standard variations, limited patent protection, minor R&D.
3	New product with proprietary technology, but may be duplicated by others, mix of standard and special features, average R&D.
4	New product with original, state-of-the-art, proprietary technology, specialized product with many adaptations, significant R&D.
5	Unique original product or system, which will obsolete existing ones, based on proprietary technology beyond the state-of-the-art, highly specialized and customized, major R&D.

added for the customer/user in terms of new or improved functions, increased benefit/cost ratios, higher reliability, and serviceability, etc.[17]

In fact, it appears that there is a positive relationship between the value of an innovation to the innovator (expressed as the stream of profits accruing to the company, entrepreneurs, and venture capitalists) and the value added of the innovation (expressed as the stream of benefits, in terms of higher performance and lower costs, accruing to the user or customer).

Extensive studies by Griliches have shown that the social benefits of an innovation are about twice the benefits accruing to the innovator.[18] Thus, generally speaking, the benefits of an innovation are split roughly fifty-fifty between the innovator and the user.

Risk Control in Planning New High-Tech Products

The development and marketing of new high-tech products are a prerequisite for success and survival. A company can expand its product line in the three orthogonal dimensions of newness.

The following guidelines should be useful in selecting, from the almost infinite possible combinations, those products or product lines that will have the highest expected value of returns to the company. To maximize expected value, it is necessary to maximize the payoff if the project is successful and minimize the three risk components related to market, function, and technology.

Maximizing Payoff

The payoff depends upon many factors, such as market size and dynamics, competition and barriers to entry, expected market share, expected return on sales and investment, expected life of the new product, and so on. All such factors are usually considered in a professionally prepared product plan. We have concluded that the payoff is positively correlated with the degree of innovation uniqueness (see table 2–5). In fact, innovation uniqueness gives to the innovating company a temporary or quasi monopoly.

Two boundary cases of pricing strategy may be considered. First, if this monopoly can be sustained by effective barriers to entry of competitors (for instance, defendable patents and highly proprietary knowledge), the innovative product will command a premium price over inferior alternative products.

Therefore, profits may be maximized by analysis of price elasticity and by judiciously achieving the optimum combination of volume versus price; that is, the combination that yields maximum profit.

Second, if the temporary monopoly cannot be sustained over a significant length of time, there is the danger of creating a price umbrella for aggressive competitors, which may be fast followers or even overtakers of the innovating company. For instance, General Electric followed and overtook EMI, the innovator, in the CAT scanner.

In this case, it may be wiser to reduce prices in parallel to (or in anticipation of) the reduction of manufacturing costs of the innovative product, according to the learning curve. Therefore, the price umbrella will be gradually lowered. Competitors, which have not yet started to ride down the learning curve, will have higher manufacturing costs and, since prices are set by the innovator, lower potential profits, if any.

This pricing strategy was successfully applied by General Motors, which passed on gradually about half of the learning curve cost reductions associated with the diesel electric locomotive to the customer. ALCO, the major producer of steam locomotives, switched to diesel when General Motors had achieved 20 percent market penetration. Too late! ALCO's costs were too high and prices, as set by General Motors, too low to survive.

Minimizing Risk

The objective is not to avoid risk totally, but rather to recognize and quantify risk; to reduce risk where feasible; and to ensure that potential rewards justify risk taking. To achieve this, each risk component must be considered separately, starting with the most serious, market risk.

Market Risk

As shown in table 2–1, market risk is lowest when offering new products to present customers through existing distribution channels and somewhat higher when using new distribution channels for present customers. Consequently, an analysis should be made of the *functional* needs (not technical needs) of the present customer base and a list drawn of possible new products, processes, and services that can be offered to these customers.

For instance, General Electric successfully developed a very profitable apparatus service business for the preventive maintenance and repair of the equipment of its customers. Initially this service was offered through existing distribution channels, the electric utility and industrial sales organization (Level 1). Later, specialized apparatus service sales channels were developed (Level 2).

Not until after significant penetration of the existing customer base had been achieved did GE attack the "new customer" segment—that is, electric utilities and major industrial plants with installed competitive equipment—

through its established apparatus service distribution channels (Level 3). This was achieved very successfully in the United States.

However, in Europe, GE had practically no established customer base and no distribution channels. Consequently, the newness level was "4," and the risk high. GE tried to reduce this risk by acquiring existing service shops in France, the United Kingdom, Spain, and so on. Results were varied and some of these service shops were later closed or sold to third parties.

All in all, GE followed a wise strategy of risk minimization, moving to the next step only after the risk of the previous step had been controlled.

Some caution is required in selecting the "customer." For our purpose, present customer is defined not only as the company being served, but also as the same "Decision Making Unit" (DMU) within that company. Failure to consider the DMU may lead to high risks, because the customer is not the same (Level 1 or 2), but different (Level 3 or 4).

For instance, in 1965–1970 General Electric found very serious difficulties in offering large data processing systems to electric utilities through its specialized computer sales channels—*apparently* same customer, new distribution channels (Level 2). The decision was then made to use the existing electric utility sales channels (Level 1). To the dismay of both computer and utility sales management, results were equally bad. Why?

The reason was that in a large segmented organization such as an electric utility, the DMUs for power apparatus and computers are entirely separate. The DMU for power apparatus is led by system, station, and operating engineers; the DMU for computers is led by financial and data processing managers.

These two DMUs normally do not interact and the purchasing function is generally too weak to integrate the purchase of such "big-ticket" items. Therefore, the "customers" were different, although the sales channels were the same. The newness level was "3" (and not "1") and consequently the risk was high.

Functional Risk

Function is what the product does for the customer (see table 2–2), not what it is (features and technology). Everything else being equal, customer resistance to change **R**, that is, to the adoption of a product with new functions, is proportional to the newness level **N**. We have here a social equivalent of Ohm's law:

$$R = KN$$

where **K** is a constant dependent on the organizational and social characteristics of the targeted industry (utilities versus semiconductor manufacturers) and individual customers (early versus late adopters). To control functional risk, it is necessary to match the *functional utility* of the product to the level of functional newness.

That is, a customer will be more willing to adopt a product if the perceived increase of functional utility is high in relation to the utility of present products.

For instance, General Electric had no difficulty in persuading hospitals to acquire CAT scanners (Level 3 of newness) because the functional utility, as perceived by the physicians (the real customers), was so high that it could easily overcome the natural resistance of the hospital administrators to acquire new, expensive, technologically sophisticated equipment.

In fact, the adoption rate of this new product was so fast that the federal government had to step in and require "certificate of need." But the tide was too strong and diffusion continued after a one-year pause. The same phenomenon is now observed with the new MRI (Magnetic Resonance Imaging) Systems, with newness Level 4.

In this case it should also be noted that General Electric's CAT and MRI diagnostic systems showed degrees "4" and "5," respectively, of innovation uniqueness (see table 2–5). Therefore, the payoff for manufacturer and user was also very high, which in turn, compensated for the functional risk.

Technological Risk

We will assume that the company has competent R&D and engineering design personnel, thoroughly knowledgeable of the state of the art, backed by an experienced project management staff. Technological risk, then, depends upon the choice of the most appropriate technology for achieving the desired functional specifications, including performance, fixed costs, operating cost, reliability, and so on.

In a technologically mature industry (for instance, steam turbine power plants) this is a rather straightforward choice. However, high-tech industries are those in which the dominant technology evolves very fast along the classical S-curve of the technological life cycle, which relates to performance/cost versus time and where new technologies are crowding in to displace the dominant technology.[19]

It has been recommended that, in consideration of the time delay required to develop and market a new product, the most advanced technology be used, which will be at or beyond the state of the art at the time of launching the new product.

Unfortunately, while the potential payoff is directly related to the newness of the technology, so is technological risk.[20] Therefore, an embryonic technology, at the beginning of the S-curve, presents high potential payoff but also high risk. Conversely, a mature technology has low payoff but also low risk.

Consequently, the choice of the optimum technology must consider this relationship, as well as expected competitive moves, the dynamics of the marketplace and barriers to entry. Experience shows that new high-tech ventures are willing to assume higher technical risk and choose the newer technologies, while established companies are more risk averse, stick to proven mature technologies, and usually switch to newer technologies too late.

Technological forecasting, technological audits, and especially the nurturing of technological gatekeepers, can be very helpful to reduce this type of technological risk.

Risk Control: The Market Introduction of New High-Tech Products

During the product planning stage, the marketing function may not be consulted or may be overruled. It is then asked to "go out and sell" a new product that presents substantial overall risk. We will now suggest ways on how the market introduction risk may be reduced by appropriate marketing and sales strategies.

Technological Risk

Obviously, marketing cannot control or change the technology embodied in the new product. However, it can monitor and attempt to control the application of the product by the customer. If the technical risk is high, marketing should suggest applications that do not stress the product to its maximum performance.

For instance, if reliability is questionable, realtime applications should be avoided or adequate back-up assured. Also, a new system may gradually be loaded to test in the field the actual capacity and "burn in" possible defects. This approach is now used for nuclear power plants. In all cases, excessive performance claims should be avoided until such performance can be proven in the field.

Functional Risk

Before launching the new product, marketing should have clearly assessed the increased functional utility for the customer, compared to available competitive products. Two boundary cases may be considered.

Innovative Product

Here the new product is truly innovative and presents to the customer new or improved functions of high utility. In this case, the primary marketing task is to convince the customer of the value of the benefits accruing from the innovative product and overcome the natural resistance to change.

This can be done by selecting demonstration sites with early adopters who will act as references and start the diffusion process. The selection of the first major customer is a key factor for success. In general, the higher the economic value of the innovation for the customer, the higher the value of this particular customer as a reference and model for subsequent adopters.

For instance, from 1957 to 1960 General Electric invested about $10 million to build and operate Project EHV, an experimental power transmission system that reached the world's highest voltage, 790 KV, in 1960.[21] Although the project had demonstrated the feasibility of operating at twice the maximum operating voltage of North America, 345 KV, U.S. electric utilities were reluctant to move to the next higher levels, 500 and 750 KV. They said the need for higher voltages was developing slowly and 345 KV was sufficient for the next ten years in the United States.

Fortunately, a major Canadian utility was faced with the need for bringing large amounts of power across long distances within a relatively short time. On the basis of the Project EHV results, they adopted the voltage of 765 KV. In 1967 this was (and still is) the highest operating alternating current voltage in the world.

Thereafter, General Electric's EHV transformers and other apparatus were installed in Canada and later in the United States, Europe, and the rest of the world.

"Me-Too" Product

In this case the "new" product does not offer any significant functional advantages compared to other available products. Here, the primary task of marketing is to *differentiate* the product from competition. There are many ways to achieve this differentiation, either with concrete added benefits to the "product package" (for instance, superior application engineering, service, and warranties) or through company image, system selling, and advertising.

Let's go back to General Electric's apparatus service in Europe. The prices GE charged for routine maintenance and repair of smaller motors and transformers were often higher than the prices of local competitors that operated in "garage shops" with low overhead. GE targeted only larger companies operating in several countries. Its service was differentiated by a twenty-four-hour worldwide emergency repair service (at premium prices, naturally).

In the case of "me-too" products, marketing should resist the natural tendency of selling technology in lieu of customer benefits. While new technology leads to improved customer benefits in innovative products, it is a poor differentiation factor for "me-too" products. In fact, it may increase the reluctance to change of prospective customers.

Market Risk

Market risk may be reduced by targeting present company customers (provided the DMUs are truly the same) and using already established distribution channels.

Alternatively, one may consider means of augmenting the company's experience base and reducing the newness quotient by using marketing or sales staff experienced with the new customers and new channels.

Joint ventures or other cooperative arrangements with more experienced firms also may be advantageous, provided sufficient integration is achieved between the two marketing and sales organizations.

The early adopter strategy of identifying and working with an innovative customer can be useful. By offering a lead customer significant discounts, guarantees, extra services, and so forth, one can enter a quasi partnership with an insider and gain important insights and guidance for attacking the new market.

Failing this, in-depth advance market research of the customers' functional needs and buying practices can reduce market risk, which will be inherently high.

Second Products

A particular case is the market introduction of second products, following a highly successful first product, whose success has been caused by its high level of innovation uniqueness (see table 2–5). Fairly often, the second product is a dud in the marketplace, to the dismay of the marketing and company management.

We must compare the *relative* innovation level of the second product versus the first, which has now become the *standard*. Here again, let us consider two boundary cases.

For the first case, the second product is truly innovatively different from the first and therefore presents increased functional utility to the user. If the same distribution channels are used, the risk is low and the product will be successful.

This is the case of Raster Technologies, a company started in the Rensselaer Polytechnic Institute Incubator in Troy, New York five years ago. Raster has reached $25 million in annual sales (of which 40 percent is exports) by introducing three generations of truly improved faster display systems.

For the second case, the second product is a "me-too" product with a low level of innovation. In this case, the functional utility advantage over the first product or competition is zero. The risk will be high, even if the same marketing channels are used.

This was the case of the IBM PC Jr. (the "Peanut"), which did not provide any innovative functions compared to the original PC or to competition. In addition, the PC Jr. was viewed as a toy. Therefore, it was not compatible with the *image* of IBM, a serious company selling primarily to business and industrial markets.

Innovation by definition is a continuing process. The second product, to succeed, must be more innovative than the first.

Risk Is Part of the Package

Risk is inherent in new product development and marketing. It is proportional to the degree of product newness along three orthogonal components: market, functional, and technological risk.

Management can control risk by planning and introducing to the market new *innovative* products (or systems, processes, services) rather than "me-too" products; by evaluating the three risk components and the overall risk during the product-planning stage; and by allocating resources to control the most critical risk components during the market-planning and market-introduction stages.

Notes

1. Booz-Allen and Hamilton, *Management of New Products* (New York: Booz-Allen and Hamilton, 1968). See also S.C. Johnson and C. Jones, "How to Organize for New Products," *Harvard Business Review* 35 (May-June 1975): 49–62. See also M.H. Meyer and E.B. Roberts, "New Product Strategy in Small High Technology Firms: A Pilot Study," *Management Science* 32 (July 1986): 806–821.

2. E. Von Hippel, "The Dominant Role of Users in the Scientific Instrument Innovation Process," *Research Policy* 5 (1976): 212–239.

3. A. Adams, "New Product Risk Strategy in Small Firms," *The Business Graduate* (Spring 1962): 64–67.

4. R.J. Allio and D. Sheehan, "Allocating R&D Resources Effectively," *Research Management* 27 (July-August 1984): 14–20.

5. Booz-Allen and Hamilton, *Management of New Products.*

6. E. Mansfield, "Entrepreneurship and the Management of Innovation," in

Entrepreneurship and the Outlook for America, edited by J. Backman (New York: Free Press, 1983), 81–109.

7. C.M. Crawford, "New Product Failure Rates—Facts and Fallacies," *Research Management* 10 (September 1979): 9–13.

8. R.G. Cooper, "Most New Products *Do* Succeed," *Research Management* 26 (November-December 1983): 20–25.

9. G.A. Steiner, *Top Management Planning* (New York: Macmillan, 1969), Chapter 19.

10. J.D. Aram, "Innovation via the R&D Underground," *Research Management* 16 (November 1973): 24–26.

11. E. Mansfield and S. Wagner, "Organizational and Strategic Factors Associated with Porbabilities of Success in Industrial R&D," *Journal of Business* (April 1975): 179–198. See also E. Mansfield, et al., *Research and Innovation in the Modern Corporation* (New York: Norton, 1971).

12. B. Little, "Risks in New Product Development" in *Canadian Marketing: Problems and Prospects,* edited by D. Thompson and D.S.S.R. Leighton (Toronto: John Wiley, 1972).

13. A. Adams, "New Product Risk Strategy in Small Firms."

14. R.G. Cooper, "Why New Industrial Products Fail," *Industrial Marketing Management* 4 (1975): 315–326; and *Winning the New Product Game* (Montreal, Canada: McGill University, 1976).

15. R.W. Stuart and P.A. Abetti, "Field Study of Start-up Ventures—Part II: Predicting Initial Success," in R. Ronstadt, J.A. Hornaday, R. Peterson, K.H. Vesper (eds.), *Frontiers of Entrepreneurship Research,* Babson College (Wellesley, Mass., 1986): 21–39.

16. R.W. Stuart and P.A. Abetti, "Field Study of Start-up Ventures—Part II: Predicting Initial Success." See Table 1 for a list of studies.

17. S.R. Craig, "Seeking Strategic Advantage With Technology—Focus on Customer Values!" *Long Range Planning* 19 (April 1986): 50–56.

18. Z. Griliches, "Research Costs and Social Returns," *Journal of Political Economy* 66 (1958): 419–431.

19. R.N. Foster, *Innovation—The Attacker's Advantage* (New York, N.Y., Summit Books, 1986).

20. P.A. Roussel, "Technological Maturity Proves a Valid and Important Concept," *Research Management* 29 (January-February 1984): 29–33.

21. P.A. Abetti and R.W. Stuart, "Entrepreneurship and Technology Transfer—Key Factors in the Innovation Process," in *The Art and Science of Entrepreneurship,* edited by D.L. Sexton and R.W. Smilor (Cambridge, Mass.: Ballinger, 1986): 181–210.

3
Marketing Lessons from Silicon Valley for Technology-Based Firms

Albert V. Bruno

Marketing often is an afterthought in the decision-making process of technology-based firms. This lack of marketing focus results in serious marketing mishaps that befall the typical technology-based company. These are discussed in this chapter in the form of marketing lessons. The conclusions and inferences that form the basis of these marketing lessons are the consequence of two levels of inputs: the judgments made by important Silicon Valley observers and participants, and many years of observation and study of marketing decision making in Silicon Valley.

The lessons presented here are intended to assist and guide decision makers as they contemplate critically important marketing decisions, the results of which can bring success or precipitate failure for their firm.

To place the lessons in context, it is necessary to understand how technology-based firms differ from their more traditional counterparts. The very nature of these differences often compounds the impact of improperly considered marketing decisions. Technology-based companies are primarily populated by engineers and scientists. Thus, key decisions in these companies are made by individuals whose thought processes are both enhanced and limited by their training—enhanced in the sense that this training is rigorous and analytical; limited in the sense that technology is often perceived to be infallible and that marketing should conform to technology and to the formulas and programmability of science and engineering. Resource allocation is also different: research and development and engineering tend to receive a much larger share of the resource pie with often a consequent reduction in manufacturing and in marketing's traditional commitments. Product life cycles in high-technology industries are considerably shorter. Hewlett-Packard reported for 1985 that less than 25 percent of its sales came from products that had been in place for more than five years, whereas 40 percent of its sales came from products developed in the previous two years and 60 percent of its sales came from products developed in the previous three years. Of course, a heavy research and development commitment must be made; during that period Hewlett-Packard spent more than $500 million on research and development.

Change is a way of life in Silicon Valley—in products, people, competitors, and competitive structures. Shortly after his arrival at Apple Computer from Pepsico, John Sculley was asked about the most surprising difference between the two companies. He replied that at Pepsico, it took two years to redesign a label, whereas in one-fourth that time at Apple, he had redesigned the entire product line. Unpredictable change is inherently risky; risk and change bring about volatility. Volatility can lead to disaster. Obviously, the stakes are higher. Founders of technology-based firms have significant impact on the culture of their organization. The founder is often still with the company in some capacity; or if not, his or her ghost tends to haunt the establishment. Finally, organizational charts are not neat and tidy in high-technology companies. Distinctions are often blurred between marketing and research and development. Engineering and manufacturing may be indistinguishable at given points in time. For example, it is not atypical to see engineering manufacture the first few units of a new product, partly because of the nature of the technology and partly because of predecessor relationships. Where the latter comes into play is in the dilemma faced by some companies as to where to place the odd founder whose initial role in the company is no longer appropriate. Sometimes power in the organization is imbalanced because of founder's role: in a Silicon Valley microcomputer manufacturing company, the vice president of research and development is considerably more influential than the other vice presidents, even though at least one vice president, the vice-president of marketing, had a twenty-year head start in experience. The reason is because the vice president of research and development owns 47 percent of the company. In another microcomputer manufacturer in Silicon Valley, a unique, unwieldy, and ineffective position—executive vice-president of marketing—was created that reports directly to the board of directors of the company. The reason? One of the founders would not report directly to the CEO of that organization.

While it is obvious that technology is the key element in the business strategy of technology-based companies, it must be in balance with marketing to be effective. As Henry Riggs recently pointed out,[1] marketing and technology should coexist in dynamic tension; one should not always be dominant. Unfortunately, technology is usually dominant, the consequence of which is that marketing often is what you do to sell the products that have been produced." With this as a backdrop, we will now examine the emerging marketing lessons.

Research Methodology

This research study represents a departure from traditional studies of this type. Rather than focusing on the firms themselves, significant industry observers and participants were asked to identify the most important marketing "lessons" that should be learned by high-technology firms participating in rapidly changing markets. The process by which these "lessons" were assembled and synthesized was as follows:

1. Thirty industry experts from Silicon Valley, including journalists, executives, and consultants, were contacted and asked to participate.

2. Twenty-one experts agreed to provide their most important marketing "lessons."

3. Sixty-three "lessons" were received, of which fifty-four were exclusively marketing and therefore relevant.

4. The fifty-four lessons were reduced to ten independent lessons.

5. The ten lessons were sent to all participants to invite their comments; the comments were reflected in the revised lessons.

Lesson I: The Bent Arrow Theory of Marketing[2]

When the typical technology-based firm introduces an innovative product into the marketplace, it can expect that over a surprisingly short period of time, the competitive environment confronting the product will be such that the product will become a commodity. This happens for a number of reasons, not the least of which is that competitors with similar technological capabilities can copy the product, often at a better price. It becomes the responsibility of marketing to bend the arrow back; that is, to make the product back into an innovation as perceived by the marketplace to avoid the ultimately destructive commodity, price-oriented competition. The mainframe computer business in the later 1960s is the classic example. Virtually any product introduced at that time required customers to undertake a significant reprogramming effort to switch from their current computers to the new product. IBM's notion of a family of computers that would be compatible essentially took a commodity situation and bent it back into an innovative one. IBM, with the System 360, had the only computer product line with identical programming spanning a wide range of performance. Consider the situation confronting Amdahl and National Advanced Systems in today's mainframe market. Both companies are "plug-compatible" vendors selling large mainframe computers and storage systems that replace IBM machines and use the same software. Both companies have had significant price and performance advantages over IBM, allowing them to take sales away from IBM. Of about 1,300 large IBM-compatible computers shipped last year, according to Dataquest, Amdahl raised its share of units from 10 percent to 13 percent while NAS stayed at 9 percent. IBM's share declined from 81 percent to 78 percent. In January 1987, IBM unleased a counter attack; the enhanced versions of IBM's 3090 class of computers, priced at $1.65 million to $11.5 million are said to provide 28 percent to 36 percent more performance for the price as existing competitive models. Analysts expect that IBM will force price cuts and owner margins on both vendors. In this way, IBM will take its traditional approach to dealing with competitors. IBM's counterattacks have repeatedly caused dramatic swings in earnings and helped put lesser competitors out of the market or into bankruptcy court. Among the major victims are Storage Technology, Magnuson Computer Systems, Two-Pi Corp., Trilogy Systems, and Memorex. It is tough to compete in a commodity market with a deep-pocket competitor that is capable of bending the arrow back.

A story called "The Parable of Three Christmases"[3] told by industry insiders in reference to Silicon Valley's less than successful entrances into the consumer watch and calculator businesses illustrates the lesson. During the first Christmas, a few participating companies found themselves in a position where demand was tremendously higher than supply; they were successful in selling all they produced at a good profit. Encouraged by the obvious success of the early entrants, a number of other competitors entered the market between the first and second Christmas. As the second Christmas approached, demand and supply came more into balance. While those that sold their products did well, they were not as successful on a per-unit gross margin basis because of price declines as were the competitors in the first Christmas. Between the second and third Christmas, a large number of manufacturers who could emulate the product entered the marketplace, many with comparable products at a much lower price. During the course of the summer and fall, the competitors not doing well increased in proportion to the total number of competitors in the market. Several decided to exit the business by dumping their inventories on the market during the third Christmas season. Those firms that did not formulate a strategic marketing plan designed to differentiate themselves from their competitors by adding innovative product features for that third Christmas were wiped out of the business by those firms that dumped their inventories and exited. The point of this story is that almost all technology-based products become a commodity, some more rapidly than others. Firms that are not in a position to bend the arrow back through marketing find themselves at the mercy of their competitors in the commodity marketplace.

Lesson II: The Rook Theory[4]

On the chessboard, the rook can only move vertically or horizontally. This limitation should be applied to the area of new product introduction for technology-based firms. Thus, a company should move vertically or horizontally but not in both directions at once. For example, figure 3–1 shows distribution choices on the horizontal axis and alternative production capabilities on the vertical axis. The rook theory holds that it is possible to expand parallel channels of distribution while holding manufacturing constant. It is also possible to hold distribution constant and develop a new product by expanding production capabilities. What is not a good idea, however, is to enter into a parallel channel of distribution while contemplating a new production capability. This is akin to starting a new business, and even very large and successful firms such as Digital Equipment Corporation and Xerox have found this to be very difficult. DEC, for example, sought without much success to enter the microcomputer business while opening retail stores. One way to deal with this problem is to subcontract or to arrange a joint venture. Nike, the successful running shoe company, subcontracts its selling to trade representatives and its manufacturing through contracts in the Far East. Manufacturing is managed by retaining quality control; Nike does its own marketing and research and development.

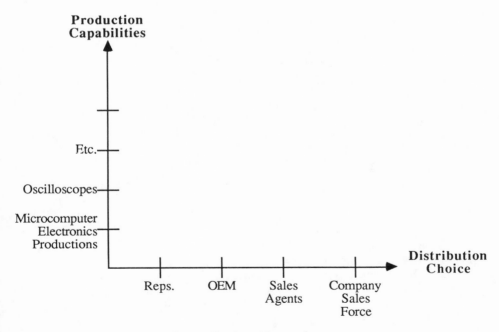

Figure 3–1. Product and Distribution Alternatives

Lesson III: Premature Marketing

Premature announcements of new products have occurred frequently in the last few years, with modest success. Many of the smaller rapidly growing firms in Silicon Valley indulging in this tactic are finding that it does not effectively preempt competition. In fact, it often preempts the firm's current product in the marketplace. Moreover, premature marketing announcements raise the expectations of potential users to a height that often cannot be satisfied when the product finally is introduced. Osborne Computer Company is the classic example of the devastative impact of premature marketing announcements. The Osborne I was enjoying considerable success when Osborne prematurely announced its Executive Line. The result was that sales of the Osborne I dropped precipitously as the market awaited the Executive Line, which never arrived. It is very challenging to deliver exactly on time the features and price/performance relationships orginally described in a premature marketing announcement.

Lesson IV: The Market Does Not Lie

Engineers and scientists employed by high technology firms have an inclination to design products that attract the admiration of their technical peers in other companies often to the detriment of the needs of the marketplace. The dynamic tension between market and technology and the lesson that the customer's needs and expectations must be satisfied become explicit in a situation of this type. Take the recent competitive environment for microprocessors, for example. A rank ordering based on technical superiority might have begun with National's 16000 down to Intel's 8086. Yet a reverse ordering from a market share point of view was actually the case: The 8086 had the largest share of the market; the 16000 had the smallest share. In the current 32-bit microprocessor market, the performance battle is between Motorola with its 68030, and Intel with its 80386 unit. The market reached $700 million in 1986 with the 32-bit chips accounting for 4 percent of it. By 1990, the market was expected to reach $2.9 billion and 32-bit chips were projected to account for 23 percent of it, according to Electronic Trend Publications. Clearly, it is important to understand the nature of the buying decision if you are one of the dozens of microprocessor manufactures competing in this marketplace. Consider the case of Grid Computer. Grid introduced a very high performance laptop computer but it failed because the market did not need all the features designed into the product at the relatively expensive selling price necessary to cover the costs of developing the features. In these and similar cases, the market does not lie. The buyer will decide based on his or her particular point on the price performance curve rather than selecting the product that provides the best performance regardless of price.

Lesson V: Sustaining Competitive Advantage Is a Marketing Imperative

There is no more important concept than competitive advantage for those making marketing strategy decisions. Competitive advantage might be defined as seeking to gain an edge over competition at a reasonable price. Many companies think they have competitive advantage, but few do. For a comprehensive discussion of competitive advantage, see Porter.[5]

In Silicon Valley, competitive advantage has been perceived to be technology driven, or in some cases market driven, but these are not the only choices. Some companies have been successful deploying other forms of competitive advantage. Texas Instruments' manufacturing capabilities have certainly made a reputation over the years for competitive advantage. IBM maintains competitive advantage through both its financial capabilities to stay in the marketplace as well as its organizational and human resource stategies. A great majority of the characteristics of the "excellent" companies as defined by Peters and Waterman[6] are human-resource oriented.

If a firm has a few ways of achieving competitive advantage and these provide little edge, the company should adopt a cost reduction mode, try to squeeze cash out, and exit where possible. If there are a number of competitive advantages to be gained but little edge once obtained, the company might seek to minimize investment, get returns up, and hold position. When large advantages are possible, but there are few ways of obtaining them, and the company is a "have," it might attack by cutting prices. If it is a "have-not," it ought to pull back. When specializing, where there are many ways to gain advantages and advantages are large, managing the position is most sensible. One excellent source of competitive advantage is product quality. The Japanese have made this a major focus of their strategy in the 1980s. It can be used as an effective marketing device, and more importantly, competitors cannot use it against you. One might consider the cost deferential between a quality control objective of even 1 percent defects versus the cost of zero defects.

Lesson VI: In-Line Delivery Systems

The basic lesson here is that the customer, products, and distribution channels all must be in line for a product offering to be successful. This means the targeted customer must find the channel to be compatible with the product. Selling a $20,000 multiuser system with all its complexities through a ComputerLand outlet will not work unless the potential user is motivated to perceive ComputerLand as a likely source for these systems. Unfortunately, there is a substantial difference between concept and practice in Silicon Valley. The point to recognize is that the user is paramount. The product sold to the user must match up with his or her needs and expectations. This is also just as true for the delivery system.

Lesson VII: Experience Is a Statistical Law, Not a Natural Law

The applicability of the experience curve has come under question in the late 1980s. By applying the experience curve, a firm has the opportunity to reduce costs by doubling the number of units sold; however, many firms have discovered that doubling volume does not automatically reduce costs. For example, in Silicon Valley, instead of accumulating management experience over a period of time such as five years, firms often accumulate one year of experience for five different management teams, as management teams frequently turn over. Thus, the successful application of the experience curve is not inevitable, but must be carefully managed.

A number of firms that do not subscribe to the experience curve have found a way to successfully deal with competitors. For example, when Intel contemplated the marketing strategy for the introduction of succeeding generations of its microchip product line, it had two basic alternatives: to reduce the price and sell the current product as a commodity or to set the price at the same level as the previous capacity level and sell at doubling of capacity. That second choice happens to be more advantageous than riding the commodity curve downward.

Lesson VIII: More Businesses Die from Indigestion Than Starvation

A study that has been ongoing at Santa Clara University for almost twenty years relates to the issues of how firms evolve over time. The focus is 250 technology-based firms started in Silicon Valley between 1960 and 1963.[7] See figure 3–2. In 1969, when the median firm was seven years old, only 5 percent had failed, an amazingly low failure rate relative to the general failure rate of U.S. firms, which approaches 50 percent in the first two years of operation. The acquisition rate was only 20 percent, primarily explained by other companies' acquiring a highly desirable technical team. At the time of the next update in 1973, the discontinuance rate reached about 25 percent

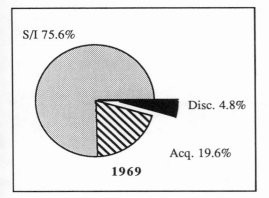

S/I 75.6%

Disc. 4.8%

Acq. 19.6%

1969

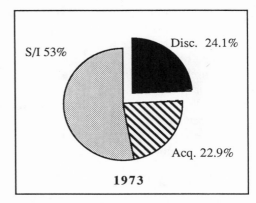

S/I 53%

Disc. 24.1%

Acq. 22.9%

1973

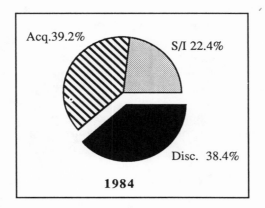

Acq.39.2%

S/I 22.4%

Disc. 38.4%

1984

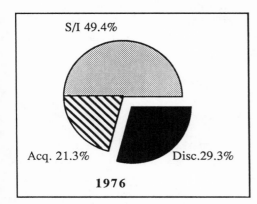

S/I 49.4%

Acq. 21.3%

Disc.29.3%

1976

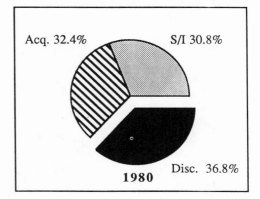

Acq. 32.4%

S/I 30.8%

Disc. 36.8%

1980

Figure 3–2. Comparative Status of Firms (1969-1984)

of the population of 250 firms and the acquisition rate was about 23 percent. In the next update in 1976, the failure rate increased and so did the acquisition rate to the disadvantage of the surviving independent firms (now only one-third of the sample). The 1984 update provided disconcerting news: The number of failures approaches 41 percent of the firms. At this stage of their existence, these firms are obviously well past their first product and probably well past their first technology; in some cases, the firms may be involved in a breadth of products and markets; that is, beyond the capacity of the firm to service. The organization either must seek acquisition or merger or some other remedy, such as Chapter 11 bankruptcy. Memorex, for example, one of the firms in our study, was doing very well until it broadened its base and got involved in too many business. The acquisition of Memorex by what is now Unisys was the consequence.

Lesson IX: Overly Managerial and Undermanaged

Once a firm reaches a certain size, it often maintains organizational trappings, such as elaborate strategic planning systems, frequent off-site management retreats, employee recreational facilities, sabbatical arrangements, company-paid educational opportunities, and the like. These manifestations of size are certainly not inappropriate in and of themselves; however, if they are replacing or hindering crucial thinking about the marketplace and how to compete to be successful, these trappings will obscure the fundamental issues and become a liability. Hayes and Abernathy in their award-winning article "Managing Our Way to Economic Decline," cited as a major finding the overly analytical managers who are more concerned with asset redeployment than running the business.[8] Our nation's business schools are guilty of perpetuating the myth of the "professional manager." There is more to stepping into a job and claiming to understand the dynamics of it: success may not necessarily require forty years of experience but it usually takes more than two years in an MBA program.

Lesson X: Growing Ventures Can Anticipate Marketing Stages

As firms grow, they pass through a four-stage marketing development process (see table 3–1). In the initial stages, entrepreneurs often sell customized products to friends and contacts. Eventually, they must exploit a larger marketplace, build appropriate internal communications and infrastructure, and diversify. Companies that successfully negotiate the subsequent stages wind up with well-organized marketing departments that effectively oversee sales,

Table 3–1
The Evolution of the Marketing Function

| | Stage 1 | Stage 2 | Stage 3 | Stage 4 |
	Entrepreneurial	Opportunistic	Responsive	Diversified
Marketing Strategy	Market niche	Market penetration	Product-market development	New business development
Marketing Organization	Informal flexible	Sales management	Product-market management	Corporate and divisional levels
Marketing Goals	Credibility in the marketplace	Sales volume	Customer satisfaction	Product life cycle and portfolio management
Critical Success Factors	A little help from your friends	Production economics	Function coordination	Entrepreneurship and innovation

marketing research, and other functions. Those that do not are often faced with no viable product development activities or with reacting to the problems at various stages haphazardly. A firm can and should explicitly consider how to successfully transit from one stage to another.[9]

Summary

This chapter examined the marketing problems confronting technology-based firms. Marketing problems for these types of firms are exacerbated by the features that distinguish technology-based companies from their more traditional counterparts. A number of such distinctions were identified: Key decision makers, including top management, are engineers and scientists; resource allocation is unique with research and development receiving a greater share of the pie; product life cycles are shorter; change is a way of life; founder-dominated corporate cultures are different. Ten marketing lessons were identified and their implications discussed. These ten topics included the bent arrow theory of marketing; the rook theory of marketing; premature marketing; the market does not lie; sustaining competitive advantage is a marketing imperative; in line delivery systems; experience is a statistical law, not a natural law; more businesses die from indigestion than starvation; overly managerial and undermanaged; and growing ventures and anticipate marketing stages.

Notes

1. For an excellent and more comprehensive discussion of strategy development in high-technology companies, see Henry E. Riggs, *Managing High Technology Companies,* Belmont, Calif.: Lifetime Learning Publications, 1983.

2. Bob Davis, formerly professor of marketing at Stanford University and now vice-president of marketing of Nike Shoes, coined the phrase.

3. Jim Unruh, now executive vice-president of Unisys and formerly with Honeywell Fairchild and Memorex, first told the "Parable of Three Christmases."

4. Richard M. White, Jr., conceived the chessboard analogy and describes it on pp. 255–256 of *The Entrepreneur's Manual,* Chilton, 1972.

5. Michael E. Porter, *Competitive Advantage,* New York: The Free Press, 1985.

6. Thomas J. Peters, Robert H. Waterman, *In Search of Excellence,* New York: Harper and Row, 1982.

7. For the most recent published discussion of these firms, see Bruno, Albert V., Joel K. Leidecker, and Joseph W. Harder, "Why Firms Fail: Patterns of Discontinuance Among Silicon Valley High-Technology Firms," *Business Horizons,* March-April 1987.

8. See William J. Abernathy and Robert H. Hayes, "Managing Our Way to Economic Decline," *Harvard Business Review,* 58 (1980), 67–77.

9. Tyebjee, Tyzoon T., Shelby H. McIntyre, and Albert V. Bruno, "Growing Ventures Can Anticipate Marketing Stages," *Harvard Business Review,* January-February 1983.

4
America's Fastest-Growing Company: Compaq's Market Creation Strategy

Rod Canion

Ompaq Computer Corporation has grown from a completely unknown start-up to become a leader in the portable computer marketplace in only five years. In the process, we have learned some lessons about running a technology-based business in today's intensely competitive marketing environment. The most important of these is bringing high technology to the marketplace and requires more innovations in business and marketing methods than it does in technology itself.

Founded in February 1982, Compaq recorded revenues of $111 million in 1983, $329 million in 1984, $504 million in 1985, $625 million in 1986, $1.2 billion in 1987 and $2.1 billion in 1988 to become the fastest growing company in the history of American business, according to *Fortune* 500 rankings. During the same period, Compaq grew from three employees to more than two thousand. How? By being entrepreneurial. By innovating not only in technology, but especially in how we take that technology to the market.

In the computer industry, a constant question is: "Why does the world need another computer company?" When Compaq started, there were more than 150 companies producing personal computers. Not only was I often asked that question by others, but I also asked that question of myself on the day we decided to produce the Compaq computer—the product around which we founded the company.

The personal computer industry has its unique characteristics, including fast growth for some companies and significant problems for so many others. Most companies have numerous unfulfilled promises that result in losses and even bankruptcies. A few—really only a handful—have had success. To be precise, only three companies have been able to consistently penetrate the primary distribution channel for personal computers, which is the computer dealer. And only two of those have been able to consistently penetrate the business market. The first, more than sixty years old and weighing in at just over $45 billion, is IBM. The second, five years old and a growing $700 million plus, is Compaq.

This presents an interesting question. Has the success (or failure) of companies in the personal computer industry been based on technical innovation? No. Certainly technology is important and it changes drastically. But in this market, as probably in most markets, technology has stayed within manageable standards. So what is the difference? What has Compaq done that literally dozens of other companies have not been able to do, or do as well?

Certainly we introduce technological innovation to our marketplace. But we really are known for our *marketing* innovation, for providing useful solutions to business problems. The answer sounds simple: innovative technology requires both innovative marketing and innovative business methods for its potential to be fully realized.

Three Major Beliefs

Three major beliefs drive Compaq Computer Corporation. First, the importance of marketing is equal to or greater than the importance of the technology that goes into the products themselves. This is one of the critical elements in our company. Unfortunately, this is most often the one element missing in technology-based companies. In an entrepreneurial environment, it is usually a technologist who really understands the technology and most often has an idea burning within him. As a result, the marketing component is either weak or missing.

Second, having the right philosophy, the corporate culture is absolutely critical in being successful and achieving high growth. Selecting the people and making certain they do the right things reflects the philosophy that is the foundation of a company. Corporate culture is really what the people who are there do—what they believe in. It provides direction. Then as other people come into the company, they should merge into that culture rather than just become an island out of touch with the company's philosophy.

The third belief concerns managing the scope of the business plan and capital availability. Certainly some areas require more capital than others, but that is not really what I am talking about. Within any area, if one understands the scope or the potential scope of a business, and can match that scope with available capital, the company then can be successful. A firm does not have to make $100 million the first year to be successful. Five million dollars the first year for Compaq would have been a great year, as long as we had planned for that and had matched it with our capital availability.

The history of Compaq Computer Corporation is really a case study of how these particular beliefs or concepts were followed. They evolved as the result of the experience we gained in marketing our products and managing the growth of the company—of seeing what worked and what did not.

Many critical events occurred and many strategic decisions were made in our first twelve months. The company was formed in February 1982. In that year, we set the foundation for what was to follow. A number of decisions were contrary to the conventional wisdom of the time. Those kinds of decisions really test management's abilities.

Compaq began with three cofounders who provided a well-balanced team. The team was planned, but it also really worked out on its own in some ways. I was an engineer by training, with some business management experience as well as some systems-oriented experience. Jim Harris was an engineering manager and a very strong engineer, who brought the Research and Development capability. Bill Murto, whose background was in product planning and marketing, brought that experience to the venture. We fit together very well. We had worked together previously so we knew we could operate well as a team. There was, as a result, a great deal of mutual respect.

Looking for Needs

When we began looking for an idea around which to build the company, the furthest thing from my mind was that we would actually build a personal computer. After all, did the world really need another computer company? We felt we would do something in the PC area, so we began the process of looking for needs in the marketplace that were not being filled. And that is the basic approach—not looking for technologies with which you could do something creative, but looking at the marketplace and asking, "What do people really need that they are not now getting?"

We came up with a surprise to ourselves and to a lot of other people. We saw that desk-style computers were moving in the direction of bigger and more powerful machines. Portable computers were moving in the direction of smaller and less powerful machines. We could see that it was technically possible to build a portable computer equal to the power of a desktop computer. Without the other key element—software—however, that product would have gone by the wayside, and we would have gone on to something else.

Personal computers don't do anything until you put software into them. I knew from having worked with software companies in the market at the time that everyone was totally consumed with developing software for the IBM PC, which had been out for only about six months. It would have been a monumental task to convince these same software companies to develop software for our product idea. Then the answer came. Instead of fighting, we decided to join—to make our product run the software that we saw all these software companies developing. At the same time, we were not entirely sure we could do that. But the basic concept of Compaq combined the hardware idea of a full-function portable computer with the software idea of being able to run software that was being developed and would be available in the future for the IBM personal computer. That was the product idea on which we started Compaq, and on which we executed for the first two years of operation of the company.

In March 1982, our second month, we were really at full steam. We closed our first round of funding at $1.5 million. That seemed like a lot of money at the time, but we knew that would not last us very long if we were to realize Compaq's potential. We faced a tradeoff involving the equity of the company. Taking the $1.5 million was really the decision to give up control of the company from an ownership standpoint. None of the founders had enough money to last very long in this market. So we made a clear decision to go for venture capital.

As the formula works, the venture capitalists put up the money, and we put up the expertise. For $1.5 million, they got 55 percent and control of the company. That was a tough decision, but it did not take us very long to make it. We knew what we wanted to do. To realize Compaq's potential, we had to give up control. So we did. But doing so we also were very careful to select the investors who were going to have control. At that time, during the venture capital boom, it was easy to find money. Money itself was not the problem. We were looking for venture capitalists to bring much more than money to the game.

We were fortunate in finding a new venture capital partnership with two strong people from the computer industry. Ben Rosen, our chairman, had been the most respected industry analyst on the Wall Street side. His partner, L.J. Sevin, had founded Mostek in Dallas and nurtured it to a several-hundred-million-dollar company. We felt we needed to have their expertise. As we look back, that was one of the best decisions we ever made. Their experience, prestige, and established reputation helped us establish our own reputation.

Compaq's first three months were extremely busy. We developed a prototype in that time and by early June we were able to begin showing it. We also began filling in the management team by hiring additional key managers for manufacturing and sales. While working quickly on the product, we were equally busy on the marketing side. We knew marketing was just as involved and just as important a process as the development cycle.

In March, we began selecting an advertising agency. Bill Murto, one of the founders and our vice-president of marketing, began an amazing process. He approached several advertising agencies in Houston and asked them to convince him why we should go with them as an advertising agency. The gateway technology we were developing never had been heard of previously. Worse that than, we could not even tell them the product we were developing. We were extremely strict on not letting out information about the product. One major agency took him seriously. They presented to him their concept for marketing a new product in the broad range of personal computers.

Throughout that process, we were able to see that they had creative talent, and just as critically, that their team fit with our team in terms of attitude and approaches. That relationship has developed into a mutually beneficial association. We believed a public relations agency would be critical for preparing product announcements and dealing with the press. Consequently, we took the effort and time to interview agencies and look for the characteristics we needed for the right fit of personalities and expertise.

In September, we were already back to the venture capital market. We raised an additional $8.5 million by bringing in eight total investors in the second round of financing. Interestingly, a lot of the people who came to talk to us during the first round would not invest in the second round. They were not interested in our idea of balancing technology and marketing or in our understanding of this marketplace. They were really looking for a technical hook, the latest technology. Another start-up company, formed about the same time as Compaq, had developed a very-high-technology portable computer with an LCD display and 3 1/2-inch floppies. It was a technical marvel. They are now bankrupt. The company's investors were persuaded by technology rather than by the ability to market that technology.

The Turning Point

In October 1982, we started production with sixty employees and twenty-six thousand square feet of space. November was a turning point. Up to that time, nobody had ever heard of Compaq. (We had worked feverishly throughout the year to pick an appropriate and marketable name.) In November, we introduced the product, had our first showing, and announced our first dealers—Sears Business Systems and ten of the major independents. In January, we announced another dealer, ComputerLand, which is absolutely critical to the overall dealer network because they represent about 350 dealers in the United States alone. Sears and ComputerLand plowed the way for all the other dealers to join.

In addition, another critical element of shipping the product fell into place. The original business plan had called for us to ship products in January. By executing that plan and delivering what we had promised, we gained enormous credibility, not only with the investors who had backed us, but also with everyone else who was watching Compaq. Estimating wrong or shipping late to the market is not necessarily the end of the world, and at times is required to match the right product with the right market conditions. But it certainly helps credibility to execute the plan as set forth.

By March, we had raised our third round of financing of $20 million for a total of $30 million in the first year of operation. We had 250 employees and a lot more manufacturing space. Our revenues of $100 million provided a tough act to follow because it created the yardstick for measuring all future growth. With this as a base, we raised another $36 million in 1983, for a total of $66 million.

Key Success Factors

Three key elements provided the foundation for Compaq during that first year of operation. First, we developed a dealer network that went far beyond the four hundred dealers we had originally planned. Because of the success of the product and its acceptance by dealers, we acquired more than nine hundred dealers, many among the top dealers in terms of service and support.

Second, in achieving the $100 million revenue level, we had put more than fifty thousand units into the hands of users. That was critical in establishing a name and reputation. If we had shipped ten thousand or twenty thousand units, we probably would not have had enough people knowing about Compaq Computers to be positioned in 1984 to deal with the tough competition that came along that year.

Third, we continued to develop the Compaq team. We not only had built manufacturing rapidly, but also continued to add steadily to marketing, R&D and, very importantly, to the control area.

With an effective control system in place, we knew where our money was, where we were spending it and the exact extent of our inventory. In other words, we had a firm grasp of all financial aspects of the company. As a result, when growth did come, we were not scrambling around, catching up, looking for inventory or trying to figure out why our expenses overran or underran the previous month's expenses.

In 1984, Compaq fully entered the worldwide marketplace and established operations in fifteen countries. We went from just a few million dollars in advertising to more than $10 million that year, five times what we had spent the previous year. This included using major network television space to build awareness of the company. We knew we needed to get more people to know about Compaq.

In 1984, we also broadened our product line. Compaq went from being just a portable computer company to being a portable and a desktop computer firm. In other words, we entered one of the broadest product lines in the market and gained a position of market leadership.

This was essential in 1984 because we saw the window for market leadership rapidly closing. Some companies were going out of business and there was a widening gap between the few leaders and everyone else. Positions were being firmly established. The difference between the fourth- or fifth-placed company and the next one down on the list was increasingly large. We had to be in that first tier of competition in 1984 if we were going to be a market leader in 1985 and beyond.

Competing with Giants

From the very start of Compaq, one question was asked of us again and again: "What are you going to do if IBM comes up with a product like yours?" In February 1984, we could put aside hypothetical responses. Many people were predicting then that IBM's product would be the nail in Compaq's coffin. It wasn't. In fact, our sales set a new record in March 1984. Sales continued to climb and we squeezed out the IBM portable by anywhere from five to one and ten to one in the computer dealer network. The reason was we had enough time to become established.

If IBM had entered the market with their product in early 1983 or even late 1983, we probably would have moved out of the running for that first heat of the race. But they didn't. Consequently, we shipped a lot of product to market, became established, and developed a recognized name. So while people looked at the IBM portable, they kept buying Compaq.

In the second quarter of 1984, we had to confront the second most common accusation about Compaq—that we existed only to fill the shortages left by others and that whenever the shortages ended, so would Compaq. IBM followed with a major price reduction of about 20 percent on their entire product line. We did not reduce our prices.

It may have seemed a natural reaction to reduce prices and stay with or below IBM. We faced great pressure in and out of the company to follow suit. However, we moved in a counterintuitive direction. Because dealers were telling us what was going on in the market, we decided not to reduce our prices. Our products continued to grow. Had we reduced our prices, it would have negatively affected our profitability and would have hurt our reputation in marketing. It would have encouraged the thinking that Compaq was just another IBM clone, just another follower. Of course, it is not.

In addition, in the second quarter of 1984, AT&T entered the market. As if one $40 billion giant wasn't enough, we got to play with two. Even this was probably positive because it kept each giant busy with the other. Still, we had to compete with AT&T. Increasingly, fate also must play a part in marketing. AT&T introduced its desktop PC to the market on June 26 in a ballroom of a hotel in New York City. Two days later, in the same ballroom of the same hotel, Compaq introduced its first entry into the desktop marketplace.

Product Strategy

Compaq's desktop product entered the market after most of the 150 companies in the industry already had desktop computers and after most had failed with them. So, we asked the fundamental question again: "Does the world really need to have a desktop computer from Compaq?" Our product strategy is really the answer to the question. Quality and reliability are essential in our products. We innovate where possible, but the product must be truly compatible. We believed our product had to be significantly better in several areas, and at the same time competitively priced.

That sounds like a basic business-course formula. But don't overlook its simplicity. That is absolutely the foundation upon which we decided on the product, its characteristics, and finally its introduction into the marketplace. The DeskPro, while remaining fully compatible, was faster and more expandable. It had more storage alternatives and a unique internal backup capability for the hard disk. At the same time, it was competitively priced with older generations. The formula worked. By the end of 1984, we had moved ahead of AT&T in the PC market. We had become the second-best-selling 16-bit PC in the desktop arena.

Six weeks after we entered the market, IBM entered the market with their new generation, the IBM-PC AT. This was interesting for a number of marketing reasons. First, it demonstrated that Compaq did not wait for IBM. We went quickly to the market with a new generation of much faster, higher-performance personal computers. We were ahead of IBM. Second, while the IBM AT has been a very good product for the firm, it has not stopped DeskPro. DeskPro has grown and continues to do very well. Third, IBM offered a rebate in the fourth quarter of 1984, that was in fact, a price reduction. Once again, we decided not to reduce our prices. The result was a new sales record in that quarter.

Top Priorities

In 1985, our top two priorities were reversed from 1984: profitability as number on; strengthening our market position as number two. Having secured our market position in the first tier of companies, it became more important that we deliver consistent profitability to be in a solid position to raise additional capital.

We were surprised in February 1985 when we were advised by our investment bankers that the market was right for Compaq to raise additional money. We had worked to position the company to raise more money when the market would open up, but we did not think that would be as soon as it actually proved to be. We sought to raise $50 million; we raised $75 million. This was at a time when Apple announced it was shutting down factories and when IBM announced the potential for reduced earnings. Because we were ready, we decided rapidly, then executed well.

All too often in a fast-growth company, its new size impedes its former rapid movement. That is the beginning of the end, at least the end of the big opportunity. Making decisions rapidly was absolutely critical to our early success. It is not any less important today. After making an important decision, having the ability to execute it rapidly is essential. This decision and implementation process involved life or death in the beginning of the company; it is still critical today. Unless we can maintain the advantage we had as an entrepreneurial, fast-growth company, we will go right back to playing on equal terms with the much larger companies. That cannot be a winning formula for Compaq.

The third priority that developed in 1985 was to enter new markets. One area we have been looking at is the telecommunications market. We formed a subsidiary different from Compaq in a different city to explore this market. Our approach is to be neither a weak nor a slow big company, but rather a group of entrepreneurially oriented, relatively small, fast-moving companies. So we are experimenting.

Getting the right people, establishing the culture, and then taking advantage of opportunities is what successful company development is all about. In this context, marketing technology is really the name of the game, not technology itself. The key is to keep focused on what the market really needs. The product must be better than the competition is offering today or what you think it can offer in the future. Developing the right technology and marketing it effectively is more important than anything else a high-tech company can do.

Part II
Accelerating the Technology Innovation Process

5
Breaking the Barriers to Technological Innovations

S. Ram
Jagdish N. Sheth

Introduction

A recent survey conducted by Booz-Allen and Hamilton reports that business executives in all sectors of the U.S. economy believe that innovation will be the major source of corporate growth and profitability in the coming decades.[1] In fact, product and service innovations are expected to generate a third of business growth and 40 percent of business profits in the United States. Innovation is also likely to be the key success factor for developing nations such as China, Brazil, and India in their desire to industrialize their economies. New products and services will continue to be an important mechanism by which corporations cope with changing customer needs, changing governmental regulations, and increasing competition. As in the past, most of these innovations will be based on technological advances. Our objective, though, is not to praise the significant role of high-tech innovations, but to explain why corporations and customers resist innovations even though they are considered necessary and desirable. We will briefly examine the role of technology in the innovation process, identify the major corporate and customer barriers to innovations, and suggest strategies to hurdle these barriers.

The Role of Technology

To fully appreciate how technological breakthroughs have had a dramatic impact on society in general and business corporations in particular, it is necessary to review the role of technology in the innovation process. Fundamentally, technology performs two functions. First, it increases the *efficiency* of natural and manmade resources. This is achieved by enabling the

---

For a discussion of these barriers in the context of innovations in consumer durables, consumer nondurables, industrial products, and services, see Sheth, J.N. and Ram, S. "Bringing Innovation to Market: Breaking the Corporate and Customer Barriers," John Wiley & Sons Inc. (1987).

resources to perform tasks faster and to their fullest capacity. This function of technology is commonly referred to as the economy of scale. Second, technology increases the *versatility* of resources by enabling them to perform newer tasks or in newer settings or both. This property is commonly referred to as the economy of scope. The greater the efficiency and versatility of resources created by a new technology, the greater its impact on society and business.

From a historical perspective, it is easy to identify the several technological ages that have triggered an explosion of new products and services. First came the *mechanical age,* based on the scientific concepts of physics and mechanics. It dramatically increased the productivity of the agricultural sector, and led to the industrial revolution. Factories and railroads were the major developments of this period, allowing for separation in distance and time between the point of production and the point of consumption.

With the discovery of electricity, the mechanical age gave way to the *electromechanical age.* Telephones, radios, automobiles, and appliances were the innovations of this era. These innovations broadened the geographical scope of markets that could be reached. The electromechanical age was supplemented by the *chemical age,* which provided breakthroughs in chemistry and biology. Innovations in pharmaceuticals, industrial chemicals, and natural resources were developed at this time and contributed to increased productivity of the industrial society.

Today, we are witnessing similar technological breakthroughs as we shift from the chemical to the *biogenetic age* and from the electromechanical to the *electronics age.* These two new technological bases may be unparalleled, because they have the scope to generate innovations that not only enhance resource efficiency but also add resource versatility. Semiconductors, represented by the chip, are now regarded as a national strategic resource comparable to minerals and oil in the earlier technological ages. Electronic communications and computers have made feasible the instant transfer of information across vast geographical distances. Laser technology has found applications as diverse as surgery and fiber optics. Biogenetic experiments are aimed at generating such innovations as the superseed, which has its own pesticide, fertilizer, and herbicide, and super plants and super animals, which are free of genetic defects.

In short, technological advances have always enabled business to develop new products and service, which are both superior in performance and lower in cost as compared to existing alternatives. In fact, nine out of ten executives surveyed by Booz-Allen and Hamilton attributed technological progress as the primary reason for innovation in corporations.

Innovation Resistance

Ironically, as the scope for innovating increases with technological advances, so does the resistance to innovate. Innovation resistance comes from two sources. On the one hand, corporations resist innovation even though survival may be at stake. On the other hand, customers resist innovation even though it could mean improved products and services. The purpose of this chapter is to illustrate why corporations and customers resist innovations. Once sources of corporate resistance and customer resistance are identified, it is possible to design strategies to overcome resistance.

Both corporate managers and consumers, especially in industrialized nations, are favorably disposed toward innovations. They believe technology can be harnessed for the benefit of mankind. Hence, the resistance to innovations is not a *cultural* problem. The resistance arises because of numerous *structural* barriers, which paralyze the desire of corporations and customers to innovate. The more radical the innovation, the greater the structural barriers, and therefore the greater the resistance.[2]

Five corporate barriers and five customer barriers may impede the success of an innovation. We shall now examine each of these barriers, and provide solutions to overcome each barrier, illustrating with examples of technological innovations.

Corporate Barriers

Most managers are aware that corporate growth and survival depends on innovation, and are pro-innovation in their values, perceptions, and attitudes. Yet, the task of innovating often becomes difficult[3] because of one or more of five major barriers: Expertise Barrier, Operations Barrier, Resources Barrier, Regulation Barrier, and Market Access Barrier.

Expertise Barrier

As technology continues to be the primary source of innovative products and services, one would expect technological specialization to be the key to successful innovation. However, this is not the case. With the exception of giants such as Bell Laboratories and Arthur D. Little, most organizations have a high degree of technical specialization. Unfortunately, the specialized knowledge is simply not versatile enough to be used for newer products. For example, when IBM attempted to move one of its immediate areas of specialization—computers—into the copying machine business, it did not succeed. Similarly, Xerox had a hard time in the computer industry. It has been extremely difficult for many central office and PBX switch makers to shift from analog to digital switches. In fact, the recent success of Northern Telecom against such strong competitors as AT&T, ITT, and Siemens is attributed to the inability of these big firms to produce good quality switches.

The reluctance and/or inability of highly specialized companies to employ a breadth of technological expertise often leads them to introduce products based on their current technological knowledge rather than on the needs of potential customers. This tendency to generate technology-driven, and not market-driven, innovations is the most dangerous consequence of the expert barrier.[4] An innovative company must be flexible enough to change its established patterns of research and development to meet the demands of the market place, but this may be extremely difficult for a highly specialized company.

Operations Barrier

The operations barrier is closely related to the expertise barrier, and is also a consequence of overspecialization. It simply occurs farther down the line in production and assembly, rather than in research and development, as in the case of the expertise barrier. A company that is highly specialized in its technology is often specialized in its operation as well. This is but a natural result of the experience curve. Innovating in such a company often involves changes in materials procurement, manufacturing, and worker training—all of which require tremendous adaptation from well-established routines.

A recent innovation in the telecommunications industry illustrates the problems posed by the operations barrier. The Integrated Services Digital Network (ISDN) is capable of integrating several types of information over the same communication network. While the telephone is capable of switching, transporting, and signaling with voice communications, the ISDN extends these capabilities to communications with data, pictures, and text with ease. However, the ISDN imposes severe operational barriers. The ISDN operation, especially at the telephone company level, is not compatible with older technologies. It requires the development of a totally new central office switch, a new PBX system, and new desktop phone sets instead of the traditional phone sets. In addition, higher capacity transport channels may be required to fully exploit ISDN's potential, in which case it may often be desirable to replace existing copper wires with fiber optics. The ISDN is thus an innovation that creates tremendous structural changes from existing modes of operation. It must be noted that the ISDN has also created expertise barriers for corporations, since it requires a high degree of integration between computer and telecommunication technologies. Firms typically excel in one of these technologies and not the other and thus have to cope with bridging the expertise gap.

Resource Barrier

Nothing discourages a business more effectively than insufficient funds. Adequate financial resources are especially influential in determining whether an innovation will ever see the light of day. Few organizations have deep pockets and can therefore ignore money as a significant factor for innovating. The resource barrier is thus a major corporate barrier that needs to be overcome.

The height of the resource barrier is determined by the borrowing power of the corporation, or in other words, its debt-to-equity ratio. Indeed, many international business experts attribute Japan's extremely successful innovation drive to the three-to-one debt-to-equity ratio that Japanese firms enjoy, as compared to the one-to-one ratio generally observed in U.S. companies. Japanese banks are not unduly worried about interest costs because the government provides low-cost capital as an incentive for firms to enter certain target industries such as automobiles and consumer electronics. The U.S. investment bankers, on the other hand, are reluctant to cross the fifty-fifty debt-equity ratio, because they fear the margins of the products may not be adequate to cover the interest costs. In any case, the lower the borrowing power, the lower the financial capital available, and the higher the resource barriers.

In the world of high-tech, a number of telecommunication suppliers are experiencing financial difficulties as they try to shift from analog to digital technologies for central office and PBX markets. In fact, ITT recently abandoned its efforts to develop the digital switch since it did not have the funds or the inclination to invest in the project. Even the German giant, Siemens, has decided to join with GTE to minimize the financial risk.

Another high-tech innovation that has had to overcome financial barriers is the cellular mobile telephone.[5,6] Unlike the regular telephone, which requires hard wires to connect with the telephone exchange, the cellular mobile phone uses airwaves to send or receive calls. It needs special, dedicated electronic switches and low-power antennae to link up with the network, and this equipment must be set up every five to ten miles. All of this requires massive capitalization. As a result, many companies that were granted permission to operate the innovative service found themselves short of capital. They, therefore, had to form joint ventures with other organizations to gain financial stability. Within a very short history of five to six years, this industry has undergone national or regional consolidation, especially among the nonwire companies.

Regulation Barrier

Regulation can take several forms, and most industries are subject to at least one of them. The first type of regulation is self-regulation, which is normally limited to codes of business practice and business ethics as expressed by the industry or trade or by a professional association. The best examples of self-regulation come from organizations such as the American Medical Association and or the American Bar Association.

The second type is government regulation of a company's internal operations as well as market operations. Government regulators are concerned with product safety, occupational safety, antitrust violations, and unfair trade practices. Federal agents operate from agencies such as the Department of Justice, the Environmental Protection Agency, and the Federal Trade Commission to enforce government stipulations. The videotex service, for example, is an innovation that has been affected to some extent due to uncertainty about government regulation.[7,8] The videotex system consists of a large database of stored information, a digital network, and a computer terminal for the users. The terminal is hooked to an existing television set and to a regular telephone line via a modem. The terminal has a remote keypad with which to input instructions. The videotex is interactive: this means the user can send messages and receive responses to them through an intelligent terminal. The system can perform business transactions such as banking or electronic shopping, or can be used for interactive education and learning by "attending" classes at remote locations. The videotex is not a regulated technology, although the Bell local telephone companies are banned from offering it on the regulated side of their business. However, it has raised a few regulatory issues. Newspapers are, for instance, worried about the competition for classified ads and the survival, they say, of the free press. This dispute has slowed down product development. Further, the regulatory climate is uncertain, and many suppliers are unwilling to invest in this innovation before the regulatory dust settles.

A third type of regulation is limited to utility services such as water, gas, electricity, and telephones. The fundamental thrust here is rate regulation, in which prices and producers are approved by a government agency. In the case of cellular mobile phones, for example, the FCC regulates the business. Currently, two companies are allowed to operate in any one geographical area. The FCC, however, permits pricing competition between the two carriers. Similarly, following the divestiture of AT&T, the Bell operating companies were restricted from entering three lines of business: information services, long distance calls, and manufacturing telephone equipment. It is because of this restriction that the operating companies are unable to offer innovations such as electronic yellow pages and certain database and transaction services. They do not have barriers from expertise, operations, or resources, but the regulatory barrier has prevented these innovations from coming to life.

A fourth type of regulation relates to patents and trademarks. Patent rights, granted by the federal government to an innovator, can protect the innovation from better imitations or poor-second-cousin innovations that seek to exploit a commercial opportunity. Patent protection has typically played a major role in drugs and pharmaceutical, and is acquiring a role of major importance in the field of biogenetics. With the patent applications being made for controversial innovations such as clones, the onus is on federal agencies to decide which innovations are morally and socially acceptable and which are not, even before deciding whether to grant a patent.

Whatever the type of regulation encountered by an innovation, the fact remains that the more regulated an industry or company, the greater the barrier to innovation.

Market Access Barrier

Market access barrier refers, in general, to all impediments that keep innovations from reaching receptive customers. It can appear because of the lack of an adequate physical distribution system or because of very strong competitors, or even customer difficulties in switching to the innovation because of the changes demanded by the new technology. The market access barrier more often creates problems for firms with low market shares than for market leaders. The smaller the market share, the greater the barrier.

A high-tech innovation that illustrates the importance of this barrier is the digital centrex. The digital centrex is designed to compete against the new generation PBX switches to handle a variety of transactions such as intercom, party-on-hold, call transfer, three-way conference calls, and the receptionist function. The centrex system requires no special equipment at the customer's office; regular telephones are adequate. All the electronic intelligence needed is located at the central office of the telephone company. Thus, unlike the PBX systems, the centrex requires little investment up front. Yet, the centrex has met with serious market access barriers. First of all, corporate customers who have already invested in a PBX system are reluctant to give up capitalized assets at or below book value, and switch to the digital centrex. Further, the switch to the local phone company's centrex could mean reconfiguring hundreds of telephone numbers and lines, a possible nightmare. Customers also have to decide whether to rent or buy terminals as a part of the centrex package, based on investment write-offs that could be obtained from outright purchase. Finally, PBX manufacturers are giving a tough fight, and attempting to upgrade customers to a fourth-generation product with enhanced capabilities for data communication through an innovation called local area networking. If the upgraded PBX is as good as the digital centrex, customers who made the switch would feel foolish. The uncertainty created in evaluating the new technology has created a market access barrier for the digital centrex.

A summary list of the five corporate barriers that impede technological innovations is listed in figure 5–1.

1. Expertise Barrier: Technical overspecialization and technology-driven
                            innovations

2. Operations Barrier: Changes required in materials procurement, manufacturing,
                            assembly, and worker training

3. Resources Barrier: Low capital resources and borrowing power

4. Regulation Barrier: Restrictions from government or from within industry

5. Market Access Barrier: Inability to reach customers because of distribution problems
                            or competitors' strengths

Figure 5–1. Corporate Barriers to Innovation

Customer Barriers

Customer barriers to innovations are more formidable than corporate barriers. Why? Clearly because innovations cannot succeed unless the intended
customers accept them. Customers do not necessarily resist an innovation
because they dislike it. It is more likely that the innovation creates change
and structural discontinuities; that is, disruption of established life patterns
they are accustomed to. This is especially the case when an innovation is
based on a radically new technology and is not just a "new and improved"
version of the existing technology.

Customer resistance to innovations is generated by one of more of five
barriers: Usage Barrier, Value Barrier, Risk Barrier, Tradition Barrier, and
Image Barrier. We will discuss each in turn with examples of high-tech innovations.

Usage Barrier

Perhaps the most common reason for customer resistance to an innovation
is that it is not compatible with existing workflows, practices, and habits.
Innovations that require significant changes in the daily routine require a
long market development process. No wonder that even innovations such as
the television, automobile, and computer were invented years before their
successful commercialization. Even if a new technology gains customer acceptance, the next wave of innovations may face usage barriers if it demands
changes in the established routines. For instance, once IBM had successfully
introduced its hardware and software technologies, efforts by other firms to
introduce alternative hardware and software architectures met with cus-

tomer resistance. Hardly surprising, because once an organization had become comfortable with IBM systems, it did not want to spend resources on retraining its users on a new system.

Video teleconferencing is another high-tech innovation that has encountered significant usage barriers. This innovation does make an honest claim in that it allows people to meet without the stress and expense of travel. However, for the conference to be successful, the organization must plan ahead very carefully. Care must be taken to see that participants in different cities have to go to specific locations away from their offices, and the time of the meeting has to be synchronized. Documents for the meeting must be prepared and distributed ahead of time, and the flow of communication between the participants has to be managed. In addition, some managers who feel comfortable with face-to-face interaction in a meeting at corporate headquarters or at a resort town, feel uncomfortable in front of the cameras and behave as if they were on stage!

Value Barrier

The second source of customer resistance to innovations is the value barrier. The value an innovation offers to a customer is its price-performance ratio as compared to existing alternatives. Unless the innovation offers superior performance and/or a strong price incentive, it is natural that customers will not consider switching to it.

The videodisc player is an innovation that could not surmount the value barrier. RCA invested more than $600 million in this technology and hailed it as a major revolution in consumer electronics. Unfortunately, the customers disagreed, and considered the videodisc player inferior to the emerging VCR alternative. First of all, the videodisc could play, but not record, while videocassettes could do both. Second, programming for disc players was restricted to only what the disc producers could offer, thus limiting its potential. Videocassettes could be reused for different purposes. Despite these drawbacks, the videodisc might have succeeded if it had been positioned by RCA as a high-quality device for serious programming: just as the hi-fi stereo was superior to the run-of-the-mill cassette player, the videodisc would be better than the videocassette. Unfortunately, RCA did not adopt this strategy. Soon the Japanese VCR producers began to drop their prices, and RCA's product had no price value either. The value barrier for the innovation was too high, hence its failure.

Risk Barrier

The risk barrier arises because all innovations, to some extent, represent uncertainty and pose side effects that cannot be anticipated completely. Customers know there are possible risks and postpone adopting the innovation till they learn more about it.

The first type of risk perceived by a customer is economic risk. If the customer has to make a considerable amount of financial investment in the new product or service, and fears he may lose it, he is reluctant to adopt the innovation. The best examples are innovations such as personal computers and video cameras, which constantly undergo improvements with technological advances. Many interested consumers postpone their purchases because they know that if they wait, a much better product with a lower price tag will be on the market. As companies dealing in these technologies learn faster and faster, their experience curves drop rapidly. The marketplace gets improved products at lower prices. Ironically, it is improved firm efficiency that creates the economic risk.

The second type of risk is physical risk: harm to persons or property that may be inherent in the innovation. For example, innovations based on microwave technology and nuclear technology[9] have met with resistance because of possible adverse effects on the human body.

A third type of risk is performance uncertainty. A customer worries that the technology may not be fully tested and developed and therefore the new product or service may not function properly. A good illustration for this risk barrier is electronic mail. Electronic mail refers to nonvoice, two-way communication between two parties using a computer terminal and a modem. It is equivalent to mail because the sender and the receiver need not be physically present or connected to the terminal at the same time to communicate with each other. To first-time users, this new technology posed some degree of performance uncertainty. The users were afraid of sending the wrong message to the wrong party, of having messages not reach the intended party, of failure to cope with sophisticated changes in the system, and of being considered incompetent. The difficulty was further compounded by the relatively risk-free character of existing alternatives such as the telephone, mail, or face-to-face communication.

Tradition Barrier

An innovation is resisted when it requires customers to make changes in the cultural traditions established by a society. The greater the change, the higher the tradition barriers, and greater the resistance.

The videotex service, an innovation we discussed earlier, had to face this barrier. Because Americans love gadgets, new technologies, and modern ways of doing things, one would have expected no tradition barrier for the videotex. Yet, a vast percentage of the mass market is culturally resistant to high-tech innovations such as computers and view them as disruptive to life. The ISDN system is another innovation that has faced tradition barriers. Corporate users have to be trained to use a single terminal for all their voice, data, and video information. The telecommunications department and the data processing department, which are typically autonomous, need to be integrated into one new department. The reporting relationships change, as will the workflow and work practices. The ISDN thus creates severe discontinuities for the users from the established corporate procedures.

Image Barrier

Innovations acquire an identity solely from their origins: product class, industry, and country. If these associations are unfavorable as a result of stereotyped thinking, they create barriers to adoption. Image is by definition more perceptual than real. Image barriers are typically built from violated social taboos, from stigmas associated with new technologies, or from deep-seated psychological forces that may be aroused by the innovation.

An innovation that has encountered image barriers is the lifeline telephone service. The lifeline service is primarily designed for the poor who can no longer afford the rising costs for the basic telephone service. Part of the problem is caused by regulation, which has eliminated most of the subsidies from long distance calls and from business customers. The lifeline service allows each residential customer a minimum number of calls, which includes emergency calls for police, fire, and ambulance. The customer must pay for each local call beyond this minimum. In some states such as California, this service is mandated by law to ensure that no one is deprived of the basic telephone facility because they can't afford it. The basic problem, though, is that customers have to declare themselves poor to become eligible for the service. Admitting poverty has negative connotations in our society. It suggests a defeatist attitude and a lack of motivation. It creates suspicions of fraud. Using the lifeline service creates a sense of guilt and shame that has been associated with other socially subsidized services such as welfare, food stamps, and public housing. Hence, the image barrier to the innovative telephone service.

A summary list of the five customer barriers is in figure 5–2.

Breaking the Barriers

Having identified the corporate and customer barriers that thwart technological innovations, we will now identify strategies to overcome each barrier.

Crashing Corporate Barriers

Let us examine each of the five corporate barriers in turn.

1. Usage Barrier: Disruption of existing workflows, practices, and habits

2. Value Barrier: Low performance-price ratio

3. Risk Barrier: Possible monetary loss, physical damage, or performance uncertainty

4. Tradition Barrier: Going against social norms and cultural values

5. Image Barrier: Taboos, stereotyping, and negative associations

Figure 5–2. Customer Barriers to Innovation

Strategies for the Expertise Barrier

The expertise barrier is created because of technical overspecialization. There are three strategies that can be used to overcome this barrier:

Use Skunk Works. As Peters and Waterman discovered in a survey of excellent companies, one highly efficient way of encouraging innovative thinking is to organize technically talented people into an autonomous task force that is completely free from the corporation.[10] This technique succeeds because it provides creative people with the freedom to experiment outside the constraining influence of established corporate thought and behavior. Such autonomous task forces, known as skunk works, have been a prime avenue of successful innovations.

A good example is IBM's success in the personal computer business. IBM was a company revolving around mainframe computer technology and big-business customers. The PC was based on a new technology and catered to a different customer base. Hence, the development of the PC met with a lot of corporate resistance. IBM therefore created an autonomous product team and gave it the freedom to do whatever was necessary for the product to succeed. The force broke some of the respected traditions in attaining success.[11]

First, the team bought the microprocessor chip from outside the company rather than developing it within. Second, it decided to sell the product through third-party distributors, which was contrary to IBM's policy of direct sales to end users. Third, the team designed the PC to be less a system and more a set of modular components, so that customers could mix and match other manufacturers' peripherals. Finally, it encouraged small entrepreneurs to develop applications software, an unprecedented departure from IBM culture. But all these creative maneuvers of the skunk works paid off for IBM.

Form Research Alliances. It is becoming increasingly common for innovative companies to form strategic alliances with individuals and organizations that wish to share their expertise for mutual profit. Recently, Whirlpool, AT&T, RCA, General Electric, American Home Builders Association, 3M, and the AMP Corporation have formed a strategic alliance to develop a new technology called the smart house technology. The alliance is exploring the possibility of wiring homes on the inside such that the same outlet can be used interchangeably for telephone, computers, television, cable, and home appliances. Unlike skunk works, research alliances allow each company to maintain its established patterns of R&D. The alliances just allow these firms to share their expertise to develop successful innovations.

Pursue Acquisitions. Another popular way of overcoming the expertise barrier is by acquiring or merging with another organization that has the technical know-how. For example, when IBM wanted expertise in the telecommunications industry, it acquired ROLM Corporation. Unfortunately, unwise acquisitions and mergers often create problems, especially if the corporate cultures of the two newlyweds clash. This can simply generate new barriers to the innovation, rather than eliminate the expertise barrier.

Strategies for the Operations Barrier

The strategies for overcoming the operations barrier are naturally aimed at overcoming overspecialization in the production and assembly functions of a corporation.

Use Separate Operations. One effective strategy is to start a new, separate operations facility divorced from current physical facilities, workers, and even management. This alternative functions just like skunk works and permits innovative activities to flourish outside the constraining influence of established operational routine. The strategy used by IBM in personal computers is again a good example.

Use Modified Operations. A second strategy, most useful in those industries where physical plant and land resources are of a significant size and scope, is the modification of existing resources to facilitate innovation. The integration of new operational procedures must, however, be performed with extreme care to avoid lowering the efficiency of current operations. The modification can be done on an organization-wide basis, or might be focused specifically on that part of the organization where the technology has become outdated and needs to be rejuvenated. For example, telephone companies are upgrading their central offices with digital switches and their local loops with fiber optics.

Strategies for the Resource Barrier

There are three strategies available to finance an innovation and break through the resource barrier.

License Agreements. The first and perhaps the most common strategy is to license the innovation to other manufacturers. Sony Corporation learned this approach the hard way. It pioneered the VCR technology and established the beta format. Unfortunately, the beta format, which is technically superior to VHS, was not blessed by the Ministry of International Trade and Industry (MITI). Most other Japanese companies decided to manufacture and market the VHS format to make it more affordable to the mass market. As a result, Sony lost the VCR market even though it had a better quality product. Sony, however, has recently changed its strategy. Rather than take on the full responsibility for producing and marketing its 8-mm videocameras, it has decided to license the technology to more than 150 companies on a world-wide basis. Thereby, Sony has reduced its financial risk and cut down on the resources needed for market development and manufacturing.

Consortiums. A consortium is a joint venture in which several potentially competitive, yet interdependent, companies get together to develop an innovative technology. The smart house project we referred to earlier is a very good example of a consortium. The objective of this consortium is to develop a new wiring technology in homes that allows appliances, electronic goods, telephones and cable television to be accessed from the same electrical outlet. This high-tech project is too expensive for any one company to undertake. Further, the expertise of different companies is required—Whirlpool for appliances, AT&T for telephones, RCA for television, and so on. The consortium reduces the financial outlay required of each of these companies, and ensures that the capital raised is used by a research group with pooled expertise. A similar consortium, called the Consultative Committee on International Telecommunications and Telegraphs (CCITT), has been organized to develop technical standards for the ISDN system.

Venture Capital. A third approach to breaking down the resource barrier for a technological innovation is to invite venture capital. In recent years, this has become a very attractive solution to generate adequate financing. Wealthy individuals are constantly on the lookout for projects that offer lucrative returns on their capital. Unfortunately, this strategy does not always succeed. For one thing, venture capitalists are not committed to any one technology. They are investors merely interested in their financial returns, and are willing to take higher risks than banks. However, they know from experience that it is much better to spread risk over several ventures rather than invest in any one venture. This tends to limit the amount of capital a single venture may generate. Even for the limited amount of capital they invest, venture capitalists tend to impose controls that may not be beneficial to the venture.

Strategies for the Regulation Barrier

Remove the Barrier. The most radical solution is to abolish the regulation by legislation. This approach has been adopted effectively by one of the regional Bell companies, created from the AT&T divestiture, to obtain its freedom from local regulation in several states.

Change the Barrier. A second strategy is to shift the regulatory jurisdiction from one agency to another. For example, it is possible to transfer jurisdiction from the local government to the state government, from the state government to the federal government, or from the U.S. federal government to international governments. In the case of the ISDN system, in several countries such as the United States, supplying customer-premises equipment such as PBXs, telephone sets, and key systems is a deregulated and highly competitive business. But in European countries, customers prefer to get this equipment from local telephone companies along with the network services. Hence, other European firms wishing to compete in the business would have to change the barrier to conform to U.S. standards, the argument being the ISDN is after all an international network that should have uniform international standards.

Bypass the Barrier. A third strategy to bypass the regulatory barriers is reorganization. In many cases, this requires forming a holding company with the freedom to offer product or service innovations the regulated entity could not. Reorganization has been used effectively in the telecommunications industry. After the AT&T divestiture, the regional Bell companies created a separate corporation in charge of all nonregulated businesses. This was done to bypass the restriction that the regulated local exchange should not subsidize nonregulated businesses such as telephone equipment, cellular mobile phones, yellow pages advertising, private networks, maintenance and repair, and international businesses.

Strategies for the Market Access Barrier

Three strategies can be used to break through the market access barrier.

Align with Dominant Vendor. If a company has trouble reaching potential customers with its innovation, one strategy would be to join the dominant player in the industry. For example, many software and peripheral manufacturers have learned to align with IBM by making IBM compatible products or selling to IBM as an original equipment manufacturer.

Develop Own Distribution System. A second strategy for a blocked firm is to develop its own distribution system. Though a very costly approach, sometimes this is the only way to break the barrier to market access. For example, small innovative companies, which do not have their own dealer network, have started to use telemarketing and direct marketing programs to reach their customers. In effect, these companies have developed their own distribution system.

Use Marketing Pull. Finally, it is possible to jump over the access barrier by adopting a pull strategy; that is, tapping the power of the people on the other side of the barrier—customers. A pull strategy is especially important if the customer is unsure how to procure the product or service. Also, in the case of technological innovations, the customer needs reassurance on the product's performance and capabilities. This can be achieved through the power of advertising, promotion, and publicity.

A summary list of the strategies that can be used to break each of the corporate barriers in provided in figure 5–3.

Crashing Customer Barriers

We will now identify strategies to overcome each of the five customer barriers to technological innovations.

Strategies for the Usage Barrier

Three strategies are available for handling the usage barrier.

Adopt a Systems Perspective. An innovating firm needs to look at how its innovation interacts with other products used and activities performed by the customer. Looking at the whole operation, a firm can get a better idea of how the new product will fit into the existing system. For example, the use of computers has increased tremendously since the development of the modem, which allows connection to and communication over the telephone network. In other words, computer usage (data communication) coupled with telephone use (voice communication) provided greater use to the customer as a general communication system.

The Barrier	The Strategies
1. Expertise Barrier	A. Skunk works B. Research alliances C. Acquisitions
2. Operations Barrier	A. New operational facility B. Fully of selectively modified operational facility
3. Resource Barrier	A. License agreements B. Consortiums C. Venture capital
4. Regulation Barrier	A. Remove the barrier B. Change the barrier C. Bypass the barrier
5. Market Access Barrier	A. Align with dominant vendor B. Develop distribution channels C. Pull strategy

Figure 5–3. Breaking Corporate Barriers

Integration. A second strategy to customer usage resistance is to integrate the innovation into the precedent activity or product. In other words, rather than selling the product directly to the end-users, it is sold to an original equipment manufacturer. For example, cellular mobile phone manufacturers are negotiating with automobile manufacturers to incorporate their product as a standard option in automobiles, just like radios, stereo systems, or air conditioners.

Force. Sometimes the usage barrier can be overcome by making the adoption of the innovation mandatory through government legislation. This strategy is risky, and can be used only in situations where it is absolutely clear to the lawmakers that the customers will benefit from the innovation. The use of lead-free gasoline was promoted by using this strategy.

Strategies for the Value Barrier

Three strategies can be used to overcome the value barrier.

Performance. The first strategy is to have the innovation provide significant performance value over existing alternatives.[12] For example, the newer telephone sets have features such as automatic redial and memory buttons that were not available with the basic telephones. Similarly, newer personal computers have higher memory and storage capacity.

Price. A second solution is to reduce the manufacturing costs and pass on the savings to customers. For example, in the minicomputer industry price reductions through cost savings have taken place at a tremendous speed in less than a decade. The larger minicomputers produced by Digital, Prime, and Data General have all been reduced to desktop models with more features and functions, but at much lower prices compared to the older-generation machines.

Positioning. As an alternative strategy, a firm can attempt to add value by appropriate product positioning. This approach is more difficult to implement and requires thorough analysis. The innovating firm must examine existing product substitutes, and position the innovation in a niche or application where it has a strong performance-price superiority over its alternatives. For example, AT&T has always positioned itself as the premiere quality corporation in the telecommunications industry, and promotes all its innovations using this positioning. Despite the tremendous influx of competitors, AT&T has managed to retain a majority of its customer base intact, thanks to its superior positioning strategy.

Strategies for the Risk Barrier

The customer's perception of risk can be minimized using three strategies.

Free Trials. The most practical method of overcoming the risk barrier is to offer the technological innovation on a trial basis to the customer with full guarantees and reassurances. However, it is not always possible to offer trials on a limited basis. For example, installing a computer system is often an irreversible decision for a company. IBM has made life easier for its customers by promising full service support right from the time of installation through the trial period and subsequent adoption. This is a strategic variation that does more than minimize risk—it eliminates risk. The videotex service has also been offered by AT&T to customers on a trial basis in Columbus, Ohio, and the trial has significantly decreased the risk perceived by potential customers.

Testimonials. A second strategy to reduce perceived risk is to obtain endorsements and testimonials from experts. Of course, the experts must objectively evaluate the innovation before endorsing it. This strategy is used by firms in almost every industry to generate acceptance for their innovations.

Package as a System. A third strategy is to introduce the innovation as a component in a system already well accepted by the customer. The customer will not be able to evaluate the product independently. The innovation will be offered to an original equipment manufacturer, whose reputation will be used to offset any risk perceived by the customer. Offshore manufacturers of electronic goods and office equipment sell their products through well-known corporations like General Electric and Savin, thus reducing the customer's perceived risk.

Strategies for the Tradition Barrier

Three strategies are available to grapple with tradition barriers.

Understand and Respect Traditions. Every firm must try to understand the cultural traditions honored by its customers. Most new product failures in foreign soil can be attributed to ignorance of and arrogance about other cultures. The same is true even of products marketed in the U.S. market, but targeted toward difficult subcultures. Corporate managers and executives need to understand the tenacity of cultural traditions held by their customers. For example, innovations are bound to have a difficult time in Middle East countries where Islam is the dominant religion and the Moslems believe that all innovation is the work of the devil.

Market Education. A second strategy for overcoming the tradition barriers is educating customers. For example, when computers were promoted as an educational tool in United States schools and colleges, there was resistance since it went against the traditional blackboard-and-lecture method. But, thanks to the intervention of the government and the initiative of several leading universities, computer literacy is being propagated among even the disbelievers. In developing nations, technological innovations that encounter traditional barriers have always needed government encouragement and support in the form of mass propaganda to pull through.

Change Agents. The third strategy in combating traditional barriers is to employ change agents.[13] This is also referred to as opinion leadership or leading edge strategy. Change agents are the large, influential customers who endorse the innovation and encourage other customers to adopt it. IBM has consistently used its leading edge customers such as banks and brokerage and insurance firms to invest early in its new technologies and let the smaller customers be influenced by these early users. The same strategy has been used by Xerox in office automation and AT&T in the videotex service. The change agents are the right people to pursue initially because they are prone to be more innovative and find it easier to break away from tradition.

Strategies for the Image Barrier

Three strategies can be used to overcome the image barrier.

Correct the Image. The first strategy is to make fun of the image and suggest how silly it is for people to carry such stereotypes. Goldstar, the Korean multibillion-dollar corporation dealing in electronic goods, jokes about the negative image people have about Korean products, thus hoping to offset the negative customer perception.

Create an Image. The second strategy is to create a unique image for the new product or service. Steve Jobs, the founder of Apple computers, did just that. Apple is the symbol of the small innovative firm that came through with a quality innovation.

Borrow an Image. The third solution to the image barrier problem is to consciously associate the new product or service with a person or object with a positive pubic image. As discussed earlier, several foreign manufacturers are selling their electronic goods under the brand names of U.S. companies such as IBM and General Electric.

A summary list of the strategies to overcome each consumer barrier is in figure 5–4.

Conclusion

Technological innovations will continue to be the source of corporate growth and survival worldwide. New technologies tend to disrupt established routines of the corporations that try to harness them and the customers who try to benefit from them. What we have attempted to do here is to identify the five corporate barriers and five customer barriers that create structural discontinuity for high-tech innovations.* We have also suggested strategies to hurdle these barriers. With the life cycles of technologies becoming shorter and competition becoming more intense, successfully innovating has assumed great importance. To this end, we hope the corporation that makes an early start at identifying and breaking the major barriers to its innovations will have a valuable edge on its rivals.

Notes

1. Booz-Allen and Hamilton Inc., "New Product Management for the 1980s," Chicago 1982.

The Barrier	**The Strategies**
1. Usage Barrier	A. Develop a systems perspective B. Integrate into preceding activity or product C. Mandate through legislation
2. Value Barrier	A. Provide performance B. Reduce price as costs decrease C. Achieve positioning
3. Risk Barrier	A. Free trials B. Testimonials C. Package as a system
4. Tradition Barrier	A. Understand and respect traditions B. Educate the customers C. Use change agents
5. Image Barrier	A. Invent a new image B. Create a unique image C. Borrow an image

Figure 5–4. Breaking Customer Barriers

2. Steele, L., "Managers' Misconceptions About Technology," *Harvard Business Review,* Nov-Dec. 1983, 133–140.

3. Bujake, John E. Jr., "Ten Myths about New Product Development," *Research Management,* Vol 15, No. 6, 1971.

4. Foster, Richard N. "Organize for Technology Transfer," *Harvard Business Review,* Vol. 49, No. 6, 1971.

5. "Cutthroat Competition in Mobile Phones," *Fortune,* February 6, 1984.

6. "Mobile Phones: Hot New Industry," *Fortune,* August 6, 1984, 108–113.

7. "Videotex," *Advertising Age,* November 16, 1981, Sec 2: S1-S23.

8. "Videotex: Technology in Search of Markets," *Direct Marketing,* October 1982, 88.

9. "Microwave: The Next Health Hazard?" *Business Week,* December 25, 1971, 21.

10. Peters, Thomas J., and Robert H. Waterman Jr., *In Search of Excellence: Lessons from America's Best-Run Companies,* Harper and Row, New York, 1982.

11. McLean, J., "IBM to Adopt 3rd Party Financing Plan," *Business News,* April 26, 1982, 24.

12. Moran, William T., "Why New Products Fail," *Journal of Advertising Research,* Vol. 13, No. 2, April, 1973, 5-13.

13. Mancuso, Joseph R., "Why Not Create Opinion Leaders for New Product Introductions," *Journal of Marketing,* Vol. 33, July 1969.

6
Marketing to Nonexistent Markets

John K. Ryans, Jr.
William L. Shanklin

> The newer and more innovative a product is the more likely it is that the public might not appreciate it at the beginning. In 1950 our company marketed a tape recorder. Despite the fact that it was a great achievement and a technological innovation for us, at the time it looked like a toy to the general public. Nobody thought about recording speeches or using a tape recorder to learn languages. I believe that in the case of an entirely new product, a market must be created, not surveyed. Another way to say this is that a new product is the creator of a market and a new product cannot survive without the creation of a new market.
> —Akio Morita, chairman and CEO, Sony Corporation

A person perusing most marketing literature, notably academic books on marketing, is told that a company's route to prosperity is the identification and fulfillment of customer needs and desires within the company's strategic competencies or lines of business. In the vast majority of the cases, this is sound advice. A company's commitment to the development and marketing of a product, process, or service almost always should be preceded by market demand as identified by marketing research. Yet, there are notable exceptions to this general rule or guideline.

Whenever a product or service is developed in response to perceived market demand, which is normally the case, an appropriate terminology to

---

Occasionally in this article, we refer to our prior research. During the period 1982–1984, we conducted an extensive formal study among some 125 leading high-technology firms. The techniques employed in the various phases of this study included in-person and telephone interviews, questionnaires, and case studies. Augmenting this formal research were our consulting and conference presentation experiences. Since 1984, our formal and informal research activities in high-technology marketing, in multiple industries and companies, has continued, thus creating an expanded data base. The data presented in table 6–1, for example, were collected in the earlier phases of the ongoing research project.

describe the process is demand-side marketing.[2] In other words, the perceived demand triggers a company's involvement with the product. This sequence is "smart" marketing—what profits are there in developing and marketing products for which there is insufficient demand? Obviously, none.

However, demand-side marketing carried to extreme stifles innovation, entrepreneurship, or intrapreneurship (that is, entrepreneurship within a going corporation rather than a startup). There are instances when a product, service, or process is developed and marketed even though it is highly problematical whether market demand is sufficient or can be created. These cases are referred to as supply-side marketing:

> Supply-side marketing, then, refers to any instance when a product can create a market . . . in other words, a demand . . . for itself in lieu of the conventional other-way-around. Or, put another way, the product is responsible for the demand, rather than the demand being responsible for the product.[3]

Supply-side marketing is fraught with risk; the two go hand in hand. But supply-side marketing—and those entrepreneurs with the tolerance for risk and courage to see their ideas through—is responsible for the technological innovations throughout industrial history. If assorted individuals and companies had waited for customer demand to justify the need for developing and marketing a new product, many technological innovations would have been a longer time coming.

Albeit supply-side marketing is inherently high-risk, there are ways to mitigate the risk. Our extensive research among high-technology companies, which resulted in our book *Marketing High Technology*, enables us to make some suggestions along these lines.[4]

Frontier Markets

One could argue that what appear to be nonexistent markets are really situations of latent demand. However, semantical debates miss this essential point: At the time the revolutionary product or service is being conceived and developed, there is no market for it. The existence of a market is contingent upon *both* a demand and a way (an applied technology) to fulfill it.

Obviously, a comprehensive cancer or AIDS cure would have tremendous market demand, but the market for most technological breakthroughs is not so manifest. Consider several supposedly expert opinions regarding the practicality and commercial possibilities of the then-new technologies:

"What can be more palpably absurd than the prospect held out of locomotives traveling twice as fast as stagecoaches."
—*The Quarterly Review,* 1825.

"I must confess that my imagination, in spite even of spurring, refuses to see any sort of submarine doing anything but suffocating its crew and floundering at sea."
—H.G. Wells, 1901.

"The ordinary 'horseless carriage' is at present a luxury for the wealthy; and although its price will probably fall in the future, it will never, of course, come into as common use as the bicycle."
—*The Literary Digest,* 1889.

"While theoretically and technically television may be feasible, commercially and financially, I consider it an impossibility, a development of which we need waste little time dreaming."
—Lee DeForest (American inventor and pioneer in radio and TV), 1926.

"There is not the slightest indication that (nuclear) energy will ever be obtainable. It would mean that the atom would have to be shattered at will."
—Albert Einstein, 1932.

"This is the biggest fool thing we have ever done . . . The bomb will never go off, and I speak as an expert in explosives."
—Admiral William Leahy, advising President Harry Truman, 1945.

Imagine a corporate CEO assessing whether to proceed with Research and Development and commercialization of a new product or process based upon the *negative* recommendation of a Lee DeForest or an Albert Einstein. A CEO might well have said, for instance, "DeForest says the commercial and financial possibilities for this television concept are remote at best. That's good enough for me." Fortunately, for the sake of technological progress and mankind, it was not "good enough" for RCA.

Researching Pioneer Markets

For purposes of market research and planning, we have found it useful, indeed essential, to distinguish between market-driven companies that operate on the demand side and innovation-drive companies that operate on the supply side. We have explained the difference between the two as follows:

In market-driven high technology, the main directions for R&D are from marketing. R&D's reaction comes in the form of guidance on what is technically feasible and ideas from scientific circles. Formal marketing research, typical to consumer and industrial markets, is helping high-tech managers guide R&D . . . Innovation-driven high technology offers a marked contrast, as R&D provides the stimulus and marketing officials must find applications or simply sell the product. These efforts can help create new markets by applying lab breakthroughs to largely unperceived buyer needs.[5]

Naturally some companies, especially large ones, simultaneously have R&D projects that are market driven and others that are innovation driven.

Market-driven and innovation-driven R&D require different marketing research approaches and techniques. The marketing research techniques most appropriate to commercializing innovation-driven R&D fall mainly in the domain of qualitative rather than quantitative marketing research. Many of the more mathematically based methods of marketing research used in market-driven ventures (that is, more mature products and services) require an abundance of data, usually obtained from a random sample of people or firms so that statistical inferences can be made. This requirement is difficult to meet in high-tech industries, for two reasons.

First, whenever markets are being created or obsoleted rapidly by significant product breakthroughs, as they often are in innovation-driven high technology, not much is available in the way of valid historical data. As one executive puts it, "Our company doesn't put much faith in surveys based on what consumers said yesterday."

Second, data obtained from prospective buyers via traditional data collection techniques (telephone interviews, paper-and-pencil mail surveys, and the like) are of dubious value for answering questions about products based on new technologies. Our evidence is overwhelming that in many high-tech markets, consumers are too confused—sometimes even scared or intimidated are not too strong of descriptors—to offer much direction to a company.

The product-education (demonstration) opportunities and in-depth probing afforded by qualitative research, such as focus group discussions, temper these kinds of confusion/intimidation problems. Thus, it is not surprising that our research indicates that focus groups and nominal groups, in conjunction with several more sophisticated techniques, are popular means in high-tech companies for generating new product ideas and evaluating potential market demand (see table 6–1).

In addition to the marketing research techniques used in innovation-driven ventures, "who to ask" is terribly important in terms of obtaining answers that are truly helpful in making go/no-go decisions and designing "winning" marketing strategies. It is far more desirable to have opinions from ten buyer-innovator types; opinions, for instance, obtained in a two-hour focus group, than to have judgments from a random sample of one thousand people, most of whom only vaguely can conceive of a new product, its technological benefits and how well it might fit into their workstyles or lifestyles.

Another more accurate indicator is the Delphi technique. As used by a

Table 6–1
Assessment of Qualitative Techniques in Generating New Product Ideas

Rank	Usage %		Not Helpful (%)	Helpful/ Extremely (%)
1	78	Brainstorming	13	87
2	58	Focus Group	17	83
3	28	Nominal Group	14	86
4	46	Attribute Listing	26	74
5	42	Forced Relationships	38	62
6	20	Morphological Analysis	10	90

jury of six experts, this method is likely to give more correct estimates of how successful dramatically new products may be in the marketplace next year than are the predictions of the most sophisticated econometric model built that must be fed a plethora of historical or cross-sectional data.

We have found that the techniques which use innovative customers, or industrial users in new product research, and methods that use the Delphi method in forecasting are popular among *successful* high-tech companies. For example, in one of our studies, three-fourths of the firms we asked indicated that they have attempted to test new product concepts on innovators; the vast majority of them reported that their "hit rate" on predicting how successful a new product will be in the marketplace was markedly improved.

Another study we conducted, this one specifically in the robotics industry, offered even more corroboration that in-depth discussions with current or potential customers is usually the best way for a company to uncover ideas for new products as well as evaluate the feasibility or market potential of technological breakthroughs. We have found that customer-innovators (those who are typically among the first to buy new products) are an especially wealthy source of useful information about new product or application issues.

Customers Buy Benefits and Value, Not Technologies

A common error in innovation-driven ventures is that the firm concerns itself with improved technologies rather than improved customer benefits. The reality is, customers buy benefits, not technologies. The relevant question is *not* whether the new technology is superior to the existing technology. Instead, the salient query is whether the new technology provides a bundle of benefits to customers, such that they are eager to abandon the older technology.

Numerous examples can be cited in which a new technology did not catch on because customers were comfortable with the existing technology or were not willing to pay more for something technically better. Emerson was wrong; if you build a better mousetrap, people will not necessarily beat a path to your door. Take Emerson's mousetrap adage literally; from a technological standpoint, a better mousetrap can be built than the wire/wood/cheese technology that has been so enduring. Yet people like the wire and wood trap and are not likely to abandon it for something technologically better, but *more costly*.

So the key to evaluating technological breakthroughs is to ask objectively whether the new technology provides better *benefits* to the customer at a price he or she is willing to pay. Is the new technology more beneficial and a better value than the old technology it will make obsolete or render less useful? For example, most automobile customers don't care whether their cars start with "points" or electronically, just so they start reliably at an affordable price.

Think about the customer benefits of these pairs of products or services: hand-held calculator versus slide rule; television versus radio; VCRs versus purchasing tapes or going to the movies; and horseless carriage versus horse-drawn carriage. In every case, the replacement technology was obviously superior to the older technology. Still, in no instance did a vast market emerge until the *perceived value* was established in enough potential customers. For instance, it took the original Henry Ford to develop the assembly line manufacture of automobiles that broke open the market for cars. Robots will replace human labor only in cases where it can be justified from the standpoint of costs versus benefits. Because of labor costs, Third World countries may find it difficult to justify robotizing their manufacturing unless this technology can provide inordinate production advantages.

Innovation-driven ventures must meet several tests, then, if they are to hope to obsolete existing products and markets and create new ones: (1) Is the new technology more beneficial to prospective buyers in terms of improving the quality of their lifestyles or workstyles? (2) Will it generally be perceived as such? (3) Will the new technology be seen as a better value (price versus benefits) than the extant technology? If not, when will it be and what will it take in the way of persuasive marketing to make it so?

Interestingly, at this writing, growth in the in-home personal computer market has slowed considerably. Why? Because the *average* person and household does not see how the benefits of an in-home personal computer are worth the price that must be paid. Obviously, the market will eventually explode, but when? Once the technology yields enough benefits that can be communicated to those in the vast middle-American market so that they are willing to pay the price. And if history repeats itself, the price will be considerably less in inflation-adjusted, real terms than it is today. Color televisions, microwave ovens, autos, and other technological innovations followed this pattern.

Establishing Commercial Priorities

Opportunity prioritization is—or should be—the high-technology company's critical initial step toward commercialization of lab output.[6] By this we mean that, subsequent to in-depth marketing research on market potential, the firm must decide which existing or presently nonexistent "future" market(s) it is interested in pursuing. When someone says market potential, market demand, or market research, it is assumed that a relevant market has been identified and is firmly in mind. For example, the nomenclatures automobile market, microcomputer market, or soap market conjure up clear pictures to most people. But in innovation-driven high technology, this is often not the case.

Take biotechnology. The potential applications of genetic engineering are countless. A biotech company could very well decide *not* to compete in pharmaceutical/medical markets at all, although these are the applications most people think of for biotechnology R&D. A biotech company might, for instance, "create" markets in agribusiness (such as disease-free orchids) or environmental pollution cleanup (such as oil spills). Similarly, the fields of lasers, robotics, and fiber optics have many possible applications (markets) that do not exist today.

Any R&D breakthrough needs to be evaluated within the context of a formal market opportunity identification analysis. In this regard, future opportunities in known and prospective or developing industries are formally assessed. After completing this stage, a company will have identified several possible applications for its new technology, applications that will likely cut across industries. And the firm will have a sense of the future prospects for these industries. At this point, some applications or industries can be eliminated from further consideration, for various reasons.

After a thorough screening, the industries and related applications that remain all will have some potential. Realistically, however, the company will be unable to exploit fully all the potential applications that have been identified, even if all the industries appear to have exceptional prospects. However, before doing the marketing research needed to better target opportunities, it is vital for the company to do some ordering of, or establishing priorities for, the prospective industries. The cost alone of undertaking extensive marketing research necessitates first designating the industry or industries that look to offer the most profitable long-term opportunities.

Once this priority-setting task is achieved, the company can turn its attention to conducting some of the indepth marketing research, particularly qualitative research among potential customers, that we have mentioned. For example, primary (original) research can be carried out to obtain more specific information on possible demand of lasers in the clothing manufacturing business. (Lasers are used to cut patterns.) With the resulting feedback as guidance, more informed choices can be made in the actual ranking and selection of target markets.[7]

Innovation-driven companies must put the "market" in market research. That is, before market research can be undertaken effectively, it must be determined what existing or future market(s) the company has in mind. Under few circumstances should a marketing researcher be turned loose to find an application for an R&D breakthrough without some firm direction as to what market(s) the company has in mind. Otherwise, the researcher is left to flounder about. But is employing good marketing research techniques and establishing commercial priorities sufficient? No, there is an additional ingredient to consider.

Role of Entrepreneurship

Indeed, effective marketing to frontier markets requires far more than establishing commercial priorities and corporate reliance on marketing research and market analysis. It also depends on entrepreneurial spirit, the right corporate culture. Marketing research and market analysis can make supplyside marketing more calculated (informed) risk-taking, but there comes a time when management must rely on intuition and fortitude to go ahead with a project. Former ITT CEO Harold Geneen makes the point in his book *Managing,* that almost all large companies that think they are entrepreneurial are, on closer examination, really not. They may fund highly risky projects, but, even if the projects fail entirely, the amount of assets of risk is so small that the net effect on corporate earnings is virtually imperceptible. Because of their fiduciary responsibility to stockholders, most *Fortune* 1000 CEOs must take mainly trustee roles rather than entrepreneurial ones. In Geneen's view, "Betting the Company" is characteristic of startup ventures, but is inappropriate for large going concerns, unless the company is in big trouble. (Which paves the way for a Lee Iacocca, who does bet the company.) And that is precisely why the net job growth in the United States in recent years has come from small business, not from the *Fortune* 500.

Geneen is right; the entrepreneurial spirit embodied in supply-side marketing is rare in large companies. Even so, exceptions exist. There are companies wherein the culture promotes innovation and entrepreneurship. And we know for sure that corporate culture, whether it be a *Fortune 500* or a startup venture, is crucial if supply-side marketing is to have a fighting chance to succeed.

Eight Masters Listed

Fortune, with the help of respected business executives and academics, has identified what are considered to be eight "corporate masters of innovation." Although these companies (American Airlines, Apple Computer, Campbell Soup, General Electric, Intel, Merck, 3M, and Philip Morris) are all large ones, they have managed to retain a culture conducive to innovation and entrepreneurship. We believe that the basic philosophies which pervade these eight companies are essential to effective supply-side marketing, irrespective of company size.

What are some of these philosophies?

- Strong sense of corporate mission.
- Paranoia about change and competitors.
- Devotion to marketing.
- Decentralization (but not anarchy).
- Unpunished subordinates' failures (except for incompetence).
- Mandatory interdisciplinary communications (such as between marketing and R&D).[8]

We do not intend to discuss each of these here but do want to elaborate on one characteristic to point out the managerial adroitness it takes to translate an entrepreneurial corporate culture into a functioning system that truly works. A host of recent articles, books, and speeches say a company has to decentralize to encourage and promote innovation and entrepreneurship, which generally is correct. But it is one thing for a CEO to say, "Let's decentralize to promote an entrepreneurial culture," and quite another to implement it.

Take two contrasting examples. IBM has been eminently successful in separating its microcomputer business unit from the remainder of the company, philosophically and physically. In spite of a few setbacks with the PC Jr., the IBM's Boca Raton decentralization was, and is, a great success story. Another computer company, Atari, decentralized with opposite results. When James Morgan became head of Atari, he found tremendous redundancy in R&D that was sapping the ailing company of what strength it had left. When

Morgan asked why one engineer lived in Louisville, Kentucky, where Atari had no R&D facilities, he was told the engineer liked living and working there. So, if not properly managed, decentralized can degenerate into near anarchy.

Moreover, decisions about decentralization necessitate considerable thought and study. It might pay to decentralize marketing, but not R&D. For instance, some years ago, Hewlett-Packard mostly centralized its R&D in order to curb R&D redundancy within its divisions and to promote intracorporate R&D sharing.

The point is, there is a great deal of glibness today about how innovativeness and entrepreneurship must start at the top (with the CEO), then pervade the organization, and finally be stimulated through decentralization. We agree with these precepts, but with the caveat that they are hard to implement in practice. Marketing to pioneer markets (that is, supply-side marketing) requires an entrepreneurial spirit, the right corporate culture, which, in turn, must be achieved through proper incentives to employees, devotion to marketing, carefully orchestrated decentralization and the rest of the aforementioned characteristics of the "masters of innovation," which is why only a small percentage of all companies are masters of it.

We want to stress that the need for supply-side marketing should not be used by a company as an *excuse* to develop and commercialize a product, service, or process. Too frequently, companies devote time, money, and effort to products that *realistically* have little or no chance to succeed commercially. Maybe some engineer is enamored with the technology behind the product, or management lets its hope to recoup the company's investment cloud its judgment.

Supply-side marketing is risky, but it is *calculated* risk-taking. Marketing to frontier markets requires strategy and work. It demands a facilitating corporate culture and organization, a focus on potential customers and customer benefits, a healthy dose of (especially qualitative) marketing research guided by lucid market and industry priorities laid down by top management, thorough analysis based on the market research and . . . intuition and fortitude.

Notes

1. Akio Morita, "Creativity in Modern Industry," *Omni*, March 1981, 6. Reprinted by permission.

2. William L. Shanklin, "Supply-Side Marketing Can Restore Yankee Ingenuity," *Research Management*, May-June 1983, 20–25.

3. *Ibid*, 20.

4. For a more comprehensive discussion, see William L. Shanklin and John K.

Ryans, Jr., *Marketing High Technology* (Lexington, Mass.: Lexington Books, D.C. Heath and Company, 1984).

5. William L. Shanklin and John K. Ryans, Jr., "Organizing for High-Tech Marketing," *Harvard Business Review,* November-December, 1984, 164–171.

6. See for elaboration, John K. Ryans, Jr. and William L. Shanklin, "High-Tech Megatenets: 10 Principles of High Technology Market Behavior," *Business Marketing,* September 1984, 100–106; and John K. Ryans, Jr., and William L. Shanklin, "Positioning and Selecting Target Markets," *Research Management,* September-October, 1984, 28–32.

7. Shanklin and Ryans, *Marketing High Technology,* 64–69.

8. Stratford P. Sherman, "Eight Big Masters of Innovation," *Fortune,* October 15, 1984, 66–84.

7
The Marketing Challenge: Factors Affecting the Adoption of High-Technology Innovations

Peter J. LaPlaca
Girish Punj

R ecently a developer of a computer software product called Configure attempted to introduce a program that would be useful in the design and operation of the "Factory of the Future." This software—a high-technology innovation—was recognized as a significant solution for many of the roadblocks that have hampered the growth of these computer-aided manufacturing systems. However, despite extensive tests with government applications, the firm was unsuccessful in attracting even one manufacturing firm that was considering investing in major new manufacturing facilities. Why was this obvious solution to a real and widely recognized need ignored?

A leading manufacturer of ball bearings came up against another situation. The U.S. robotics industry has faced an uphill climb against stiff foreign competition. In recent years several firms have produced improved industrial robots that have proved satisfactory in a wide range of manufacturing applications. Not long ago, a ball bearing manufacturer designed some significantly improved, technologically innovative bearings that would enhance robotic performance, only to drop its efforts in this market on the basis of marketing research. This decision saved the company several millions of dollars, but the company could have saved even more if it had had this key information earlier.

The experiences of these two companies illustrate the need for the three-dimensional innovation model we will describe in these pages.

By more completely integrating technology, product, and market—the three dimensions of technological innovation—high-technology companies can increase the chances of successful adoption of innovations and have a better sense of which innovations are likely to fail. As a result, they can develop more appropriate and effective marketing strategies and tactics.

Studies of the diffusion of innovations most frequently reflect two theoretical frameworks—the adoption perspective[1] and the economic advantage perspective.[2] While the adoption perspective has been useful in understanding the diffusion process for a great many nontechnical innovations, this model has had mixed results when applied to technological innovations. Likewise, the economic advantage perspective does not fully describe all significant factors associated with technological innovations. This chapter is an extension of the framework for the understanding of the diffusion of high-technology innovations based on existing research of ordinary and technological innovations.

Investigations of innovations have usually focused on the population into which the innovation has been introduced or on the innovation itself. This chapter integrates these two streams of research by examining the process which leads to the introduction of the innovation into the market. It presents a new model of the development process for high-technology innovations along with a comprehensive system for structuring alternative diffusion models. The system is based on differing characteristics of the innovation, the technology, and the market.

Innovation Classifications

The traditional classification scheme for innovations has been to place the innovation into a continuous, dynamically continuous, or discontinuous category based upon the degree of behavioral change induced in the adoptor.[3] Under this scheme a continuous innovation may be best represented by a modification to an existing product or service which, while newly introduced into a market, does not require the users (adopters) to alter their purchase or consumption patterns. Examples would be such items as pump packaging for toothpaste or a more durable brake pad for light aircraft.

At the other extreme, discontinuous innovations are entirely new products or services that require or cause a drastic modification of the users' purchasing or consumption behavior, such as enabling the user to perform an entirely new function. Examples would include laser cutting technology for the garment industry, CAD/CAM, and Computer Integrated Manufacturing Systems, artificial fertilization of humans, or the original introduction of television to the consumer market.

Dynamically continuous innovations are those that lie between these two ends of the innovation continuum. This has been most appropriate for household innovations directed at consumer markets, but does not reflect many of the factors affecting the adoption decision for innovations based on advanced technologies such as those frequently directed to industrial and other business markets. Household innovations (also referred to as ordinary innovations) are those new products or ideas adopted by individuals or households (such as electric toothbrushes, menthol cigarettes, microwave ovens, and home computers). Many industrial innovations represent the replacing of an old technology with a new one, such as the adoption of new machines, processes, and techniques by firms for their own use[4] (such as process control equipment, the Bessemer steel process, business desktop computers, and xerographic copying machines). Because of this, the term "technological innovation" is frequently applied to new products, processes, and ideas directed at industrial and business markets.

An important factor of technological innovation is that, while both the old and the new technology can accomplish essentially the same task, the replacing technology provides distinct economic advantages over the old technology. In fact it is this emphasis on economic factors such as the relative profitability and required investment that distinguish ordinary innovations from technological ones.[5] An additional factor that distinguishes technological innovations from ordinary ones is the process of re-invention, where adoptor modifications of the innovation as introduced to the market by the inventor firm can broaden the range of useful applications of the innovation.[6]

While these technological innovations present alternative means of accomplishing existing tasks (usually more economically), some innovations go beyond this to create the ability to perform entirely new functions. These "high technology" innovations represent that segment of technology considered to be nearer the leading edge or state of the art of a particular field.[7]

High-technology innovations, therefore, represent the addition of technologies to the adoptor firm's processes to enable it to accomplish entirely new tasks. By accomplishing these new tasks, previous tasks may become unnecessary; therefore, high-technology innovations represent a redirection rather than a replacement of existing technology (speech recognition equipment, earthquake predictors, lasers, and biogenetic engineering).

Existing Diffusion Frameworks

The common approach to studying innovations (derived from household and agricultural innovation research) is the adoption perspective. This perspective focuses on the patterns of information exchange among members of the population of interest (the target market) and the rate at which this information is transmitted throughout the adopting population. This approach requires an understanding of the characteristics (that is, demographic, sociological, and psychographic factors) of the adoptors (vis-à-vis the nonadoptors) almost to the exclusion of an analysis of the innovation itself. Thus, the underlying theory behind the adoption process is that it is an imitative phenomena.[8] Therefore, the adoption of the innovation depends on the effective flow of information and the communications and learning processes.

Under this perspective, resistance to the adoption of the innovation is the result of a mismatch between the benefits provided by the innovation (as perceived by the potential innovator) and the innovator's tolerance for risk-taking behavior (the risk that the benefits may not actually be realized).

The most common method used to study technological innovations is the economic perspective. This perspective involves the determination of relative economic advantage of a new technological innovation as compared to that which it replaces. Economic advantage can be measured by means of overall cost savings to the innovator (life-cycle costing or costs incurred over a specific time period), improved productivity and throughput, and increased profitability resulting from use of the innovation. These measures are obtained by an analysis of the innovation itself rather than a study of the adopting firms. The underlying philosophy of this approach is economic in nature.

These economic factors are difficult to measure for several reasons.

- Measures of profitability must incorporate estimates of sales revenues as well as costs.

- The cost structure associated with most technological innovations is dynamic in nature as the inventor firm attempts to gain advantages of experience in production processes.

- Management philosophy toward new innovations and their inherent risk-taking behavior can cause resistance to the technological innovation.

- Many of the anticipated economic advantages associated with adoption of the innovation may not be realized within the time frame encompassed by management's decision horizon. (Or put another way, long-term benefits requiring investments today may be sacrificed for improved short-term profitability or cash flow.)

The study of high-technology innovations requires an integration and extension of the adoption and economic perspectives to more fully understand continuous and discontinuous types of innovations. This integration includes analyses of the technology upon which the innovation is based, the nature of the innovation itself and the markets it serves and results in a three-dimensional technology-product-market configuration. This extension overcomes the limitations of the adoption and economic perspectives of diffusion. Moreover, it applies to the entire spectrum of innovation categories. The broader scope of this three-dimensional approach provides a fuller description of the factors affecting market success of the innovation.

Conceptual Framework

The three dimensions that form the basis for the proposed model of innovation diffusion are the technology, the product, and the market. These dimensions are inherently continuous in nature; however, for the purpose of explanation, they will be presented as being discrete thus resulting in the 3 × 2 × 2 matrix shown in figure 7–1.

The technology dimension serves as a basis for measuring the degree of absolute newness of the innovation. It encompasses several factors:

- the existing base of technology (how long the technology has been available, the closeness to the state of the art, and the acceptance of the technology by those researching the field);
- compatibility of the technology to accepted applications in the market;
- the rate of technological change;
- the ability to provide a measure of relative advantage of the innovation.

High-technology-innovations are those innovations that:

- are closer to the state of the art of the technology in the field;
- do not have widespread acceptance by those working in the field;
- lack compatibility with accepted applications in the market;
- are experiencing high rates of technological change;
- have difficulty in providing measures of relative advantage.

Standardized innovations are those nearer the opposite end of the technological dimension. They reflect modifications of existing products or combinations of existing technologies to produce new products or services offering the features of the technologies from which they are derived.

Technological innovations are those at neither extreme. Examples of these types of innovations would include household applications of technologies that had been developed for advanced applications (such as the space program or undersea exploration) or combinations of existing technologies that result in capabilities exceeding those of the separate technologies that had been combined.

The product dimension represents the form in which the technology enters the market. In this case, it is an indicator of the newness of the product to the inventor firm. It involves such items as whether the innovation is new to the world or merely new to the firm, the degree of modification of the innovation from existing products and the amount of value added by the introducer firm. Existing product "innovations" are those new to the firm

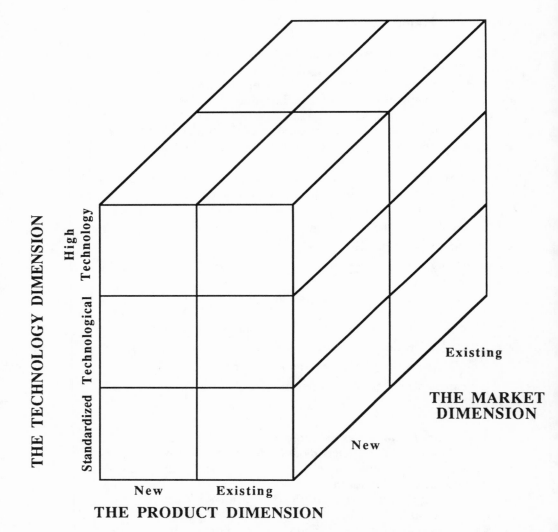

THE TECHNOLOGY DIMENSION

High — Standardized Technological Technology

THE PRODUCT DIMENSION

New   Existing

New   Existing

THE MARKET DIMENSION

Figure 7–1. Technology/Product/Market Dimensions of High-Technology
Innovations

but not new to the world, that have fewer modifications from existing products and have relatively low value added. New product innovations have the
opposite characteristics.

The market dimension looks at the population of potential adoptors
from the perspective of the inventor firm. Those potential adoptors already
being served by the firm (with different products and/or services fulfilling

different needs) comprise existing markets; other potential adoptors comprise new markets.

Inventor firms are more familiar with the needs and purchasing systems of existing markets; indeed, the assessment of unfulfilled needs of existing customers frequently is the impetus to develop a new product or service. Customer modification of existing products can easily point to new (continuous) innovations. Knowledge of purchase systems (as well as specific individuals) can greatly ease the communication of information to the proper decision makers in potential adopter organizations. Existing markets pose fewer problems of synergy with distribution methods of the firm and are more closely aligned with communication patterns and information flows familiar to the inventor firm.

New markets may require totally new distribution systems or extensive modifications in the methods used by the firm to inform potential adoptors about the innovation. It is not unusual to combine the market and product dimensions into the common concept of product/market matrix that serves as the basis for development of marketing strategies.[9] However, this approach is most suitable for applications to the firm's existing product line or for well-defined markets; it presents an insufficient framework for the strategic analysis of products still in their fetal state with indeterminable gestation periods.

This three-dimensional model can accommodate existing diffusion perspectives. For example, the adoption perspective fits into the market dimension of the model through research on communication patterns and population structures, adoptor/non-adoptor characteristics, receptivity to the innovation, and other factors related to the population of interest. In the economic perspective, the emphasis is on the innovation or the product itself. Analysis under this scheme involves economic benefits derived from product use by the potential adoptor.

Integration of these two perspectives involves studying product factors (relative advantage, complexity, trialability, compatibility, and observability) alongside market factors (communication patterns, rate of information dissemination, and adoptor innovativeness). However, product-market interactions (such as factors affecting trialability among early adoptors—for example, installation of prototype machinery or incorporation of preliminary components into the adoptor's products) are not part of the examination. The proposed three dimensional model permits analysis of two- and three-way interactions among technology, product, and market factors.

In fact, a marketing professional can initiate the model with a unidimensional investigation of market factors, improve upon it by incorporating product factors, and fully apply it to high-technology innovations by adding the technology dimension.

Developing High-Technology Products

This tripartite model is useful as a hypothetical framework for researching the high-technology diffusion process. However, additional factors and a different perspective are necessary to apply this approach to the inventor firm.

Many high-technology firms are preoccupied with the technology surrounding their product offerings. They frequently cite their inadequate knowledge of market needs, means of product evaluations, or decision-making processes as the primary reasons for their newly introduced, high-technology innovations to fall short of expected market penetration and the resulting financial performance.

High-technology firms frequently lack a systematic approach to product evaluation, development, and planning.[10] The inventor firm needs accurate information on product characteristics and features required by the prospective adoptors. This includes the exact functions to be performed, the value placed by prospective adoptors on these performances and the varying requirements of different market segments. While this requirement for accurate information is derived from a logical extension of the model, it almost never exists in actual applications. When it is present, it frequently results from reactions to a prototype or a detailed product concept test. Even so, it is at best a probabalistic estimate of the match between the firm's innovation and the market's need.

It is difficult for the high-technology firm to know what market application to pursue in developing and defining specific innovations. This is especially true if several potential applications of the basic technology appear attractive at the outset. The problem becomes more complex if the product concept is entirely new to the company and if many potential applications are outside its served markets.

Therefore, a comprehensive innovation development model based upon the three dimensions of technology, product, and market can provide a mechanism to address the problems of high-technology firms. The model (shown in figure 7–2) provides a systematic framework to:

1. Identify market and technological information to develop "go, no-go" decisions for specific product-market applications within a reasonable period of time and cost.

2. Identify information necessary for the development of a complete marketing program in terms of product development, prototype testing, overall marketing strategy, and feedback and control.

3. Show flexible and synergetic relationships between the firm's existing product base and the innovations being prepared for market.

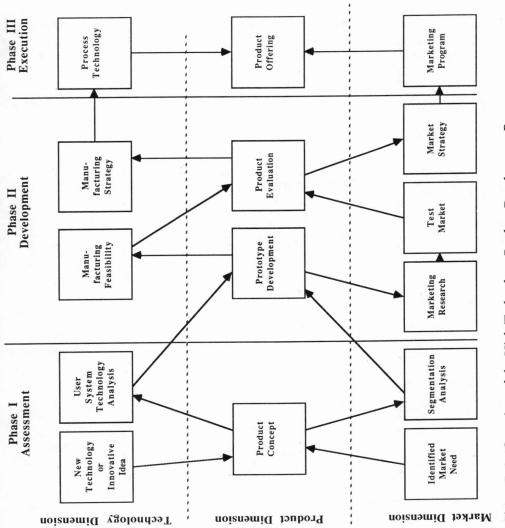

Figure 7–2. Overview of the High-Technology Product Development Process

Within a high-technology firm, the market innovation development process evolves naturally in three phases: (1) assessment, (2) development, and (3) execution (market introduction). The model presented in figures 7–2 through 7–5 structures those processes by identifying key informational needs and decisions. The end result is an innovation development process that allows management of the inventor firm to effectively assess market opportunities and risks and to integrate technological resources and marketing programs necessary for successful market entry.

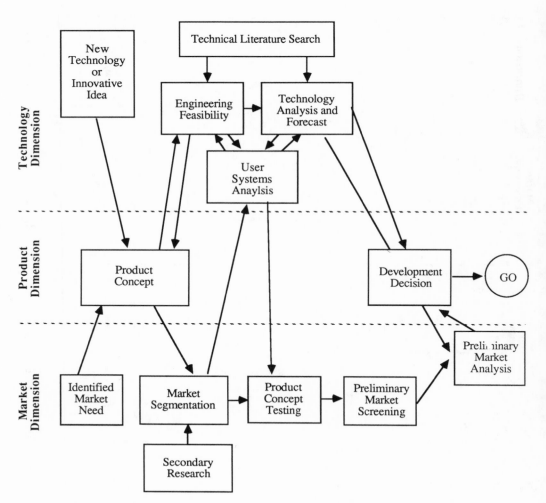

Figure 7–3. Phase I: Assessment of Product Viability in the Market

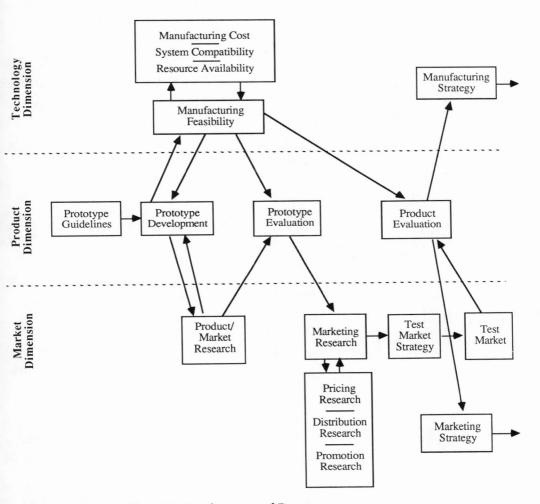

Figure 7–4. Phase II: Development of Prototypes

In each phase of the process there is an evolution and interaction among technological, product, and market dimensions. The technology evolves over the three phases:

from technological innovation to

user system technology analysis to

process or manufacturing strategy and technology.

The product develops from:

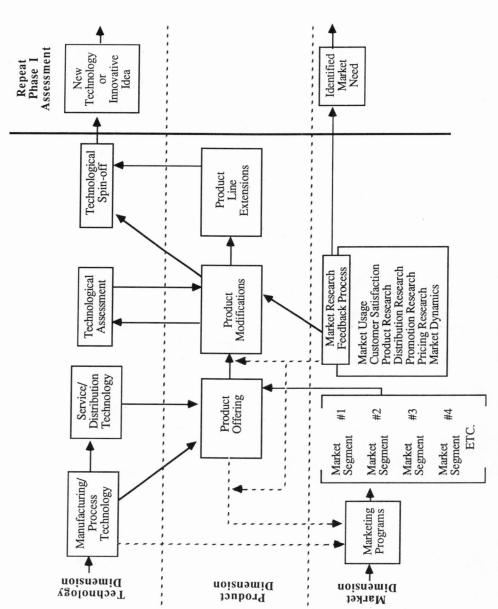

Figure 7–5. Phase III: Execution to Meet Customer Needs

a product concept
   to a prototype
      to a marketable product and an extended product line.
The marketing function evolves from:
   identification of market needs
      to market segmentation and marketing research
         to development of marketing strategies and programs.
As the technology, product, and market dimensions evolve, they continually
interact and feed back, until each is sufficiently developed to be accepted by
the target markets.

Phase I: Assessment of Product Viability in the Market

The assessment phase (see figure 7–3) provides a timely, low-cost evaluation
of the innovation's viability as a marketable product. Central to the assessment phase is the product concept. This serves as a crystallization of the
original innovative idea or proposed answer to an identified market need. It
also acts as an integrative point for engineering feasibility, analysis of potential adoptor systems, market segmentation, applications research, market
screening, and market analysis. The primary purpose of the assessment phase
is to reach a decision for further development before the project has absorbed
significant amounts of research and development resources.

The technology dimension requires three principal analyses. The first
concerns the engineering feasibility of the product concept. This provides
information on how the product should be configured to best provide the
anticipated benefits/functions and the design alternatives. The second analysis investigates the appropriateness of state-of-the-art technologies for meeting market needs as well as the rate of change for these technologies (is there
a significant chance of the proposed technology becoming obsolete before
an adequate return on the investment is obtained?). Most of these two analyses can be accomplished by means of secondary (that is, published) data or
discussions with industry experts with a modicum of engineering effort. The
third technological analysis requires an understanding of how the proposed
product concept could become part of the user's overall system. Will the
anticipated product benefit actually result, to a significant degree, when the
product is applied by the user?

For example, development of a faster piece of machinery with three
times the output of its predecessor may yield absolutely no benefit to potential adopters if there are other bottlenecks in the manufacturing process that
limit the overall throughput of the entire production system. A thorough
understanding of the possible application of the product concept by both the
producing firm and the potential adoptors in each potential market segment
is required to complete this analysis. It is important to understand both how

the product concept in use will affect the user's actual operation and the perceived risk of adopting the innovation. As described in the first case, this perception of risk can be a significant factor for individual deciders in the adopting firm.

The market dimension requires several types of analysis. One is the identification of alternative market segments. All too frequently, the technological firm investigates only one potential use of the product concept; if this market proves to be too limited to support the growth expectations of management, the project may be terminated. However, if multiple applications (at least of the basic technology) in a variety of markets are identified, there are frequently opportunities for greater cost reductions (because of the effects of shared experience and learning curve factors) in future stages of the product life cycle. While there may be a tendency to label any possible application as a possible market segment, some preliminary screening criteria (such as minimum size or growth rate) is necessary to avoid expenditures for market analyses where the potential for sales is quite limited. This preliminary market screening also will enable the firm to rank possible markets according to their sales potential for sequential investigation as the development process progresses.

After delineating the appropriate market segments, it is necessary to conduct product concept (value in use) tests in potentially attractive segments. The product concept test asks the questions, "Is this product concept viable?" and "What adjustments are necessary to enhance its marketability?" The product concept test should present the complete concept rather than separate components of the concept. In this way, management can make various trade-offs (such as reduced speed for lower operating cost) and prospective users can better visualize the product concept in use.[11] Market screening follows positive feedback from the product concept tests. This involves a quick appraisal of the potential match between the product concept and user needs. It will eliminate those market segments for which there are limited product applications and hence, limited sales potentials. Those segments that pass this initial screening face a more thorough analysis to determine current and anticipated market trends as well as a rough estimate of total market potential. This market analysis, in conjunction with the technology analysis and forecast, forms the basis for the decision to drop the project or to proceed with prototype development (Phase II). By focusing on the external environment and integrating all elements, the likelihood of success or failure quickly can become obvious.

Phase II: Development of Prototypes

Phase II (figure 7–4) expands and tests the technology, product and market dimensions of the assessment phase. The decision to proceed to this phase precipitates the development of prototype guidelines from which one or more prototypes are produced.

Management should put these prototypes through a complete manufacturing feasibility analysis, including full-scale cost projections (at multiple volume levels), tests for compatibility with present manufacturing systems (and determination of possible shared experience factors), and determination of resource availability and stability (such as raw materials and components, certainty of supply, and dependence on restricted sources).

Market segmentation progresses to customer identification and usage and delivery systems. Prototypes are field tested and necessary modifications are made. To speed up the development process, pricing, distribution, and promotional research proceed somewhat independently until the end of Phase II when they are integrated with the product for test marketing.

Test marketing helps refine previous estimates of market response and required marketing effort. At this time, the opportunity exists to conduct financial analysis, quality control studies, and other quantitative assessments of proposed marketing strategies. In addition, the overall decision analysis incorporates considerations for marketing, competitive, financial portfolio, technological, and regulatory risks.[12] The test market may result in a go-decision, recycling within Phase II for further development or an abort-stop decision. A go-decision will require comprehensive marketing and manufacturing strategies.

Phase III: Execution to Meet Customers Needs

The execution phase (figure 7–5) focuses on matching product characteristics and market delivery systems to meet customer needs. On the technology dimension, management must translate manufacturing strategies into specific process or manufacturing technologies. This occurs simultaneously with the development of detailed marketing programs for each targeted market segment. An important consideration at this juncture is the identification of probable early adopters and opinion leaders. These can be estimated from prior market behavior relative to the adoption of innovations as well as identification of specific applications that would yield high economic or other benefits from use of the product.

To control the total marketing program, management must establish a marketing monitoring system. This includes measurements of market penetration and growth, customer usage and satisfaction, and other elements of market dynamics. Additionally, management must carefully monitor the impact of the firm's product, pricing, promotional, and distribution programs for information on strategy adjustments and new product opportunities.

Continuous monitoring of customer applications (by means of the sales-and-service force) can yield useful information on customer-initiated product modifications or unmet customer needs. Product modifications should go through a technological assessment to determine the impact of the proposed changes. These modified products can be offered to other customers as alternative models of the basic product. Customer monitoring also can help identify maintenance and repair parts and services as well as related product opportunities that, subject to favorable technological assessment, also can be offered. Gradually these modified and related products will constitute a complete product line.

Product modifications sometimes can be so significantly different from existing products as to require new technologies on the application of new technologies to customer problems. In either case, a technological spinoff can result, using a venture management concept.[13] These spinoffs and identification of new market needs provide a direct linkage back to Phase I.

Notes

1. Rogers, Everett M., *Diffusion of Innovations,* The Free Press, Inc. (1962).

2. Brown, Lawrence A., *Innovation Diffusion: A New Perspective,* London, Methuen & Co. Ltd. (1981).

3. Robertson, Thomas S., *Innovative Behavior and Communications,* New York, Holt Rinehart & Winston (1971).

4. Malecki, E.J., *Innovative Diffusion Among Firms,* Ph.D. dissertation, Ohio State University, Columbus, Ohio (1975).

5. Brown, Lawrence A., *Innovation Diffusion: A New Perspective,* London, Methuen & Co. Ltd. (1981).

6. Rogers, Everett M., *Diffusion of Innovations, Third Edition,* The Free Press, Inc. (1983).

7. Rexroad, Robert A., *High Technology Marketing Management,* New York, John Wiley & Sons, Inc. (1983).

8. Rogers, Everett M., and Shoemaker, F. Floyd, *Communication of Innovations: A Cross Cultural Approach,* The Free Press, Inc. (1971).

9. Ansoff, H. Igor, *Corporate Strategy,* New York, McGraw-Hill Book Company (1965).

10. Zarecor, William D., "High Technology Product Planning," *Harvard Business Review,* January-February, 108–115 (1975).

11. Wind, Yoram; Grashof, John; and Goldhar, Joel, "Market Based Guidelines for the Design of Industrial Products," *Journal of Marketing,* July, 27–37 (1978).

12. Shah, Kiran, and LaPlaca, Peter J., "Assessing Risks in Strategic Planning," *Industrial Marketing Management,* February, 77–91 (1981).

13. Hill, Richard, and Hlavacek, James D., "The Venture Team: A New Concept in Marketing Organization," *Journal of Marketing,* July, 44–50 (1972). Also, Hlavacek, James D., and Thompson, Victor, "Bureaucracy and New Product Innovation," *Academy of Management Journal,* September, 360–372 (1973).

References

Shanklin, William L., and Ryans, John K., *Marketing High Technology,* Lexington, Mass., Lexington Books (1984).

Utterback, James M., "Innovation in Industry and the Diffusion of Technology," *Science,* February 15, 658–62 (1974).

8

Fast-Track Marketing: Stages of Growth in Ashton-Tate

Edward M. Esber, Jr.
Michael Stone

Since the company was founded in 1981, Ashton-Tate has been on the fast track. The company has annual revenues of more than $200 million, employs more than 1,100 people and has a commanding market share of a growing segment of the microcomputer industry. Such phenomenal growth has been neither simple nor easy. The company has had to weather a major recession in the U.S. economy, develop an organizational infrastructure to support explosive growth, and most importantly, adapt management strategies to meet new marketing challenges.

Company Profile

Ashton-Tate has four main software product lines for the microcomputer.

The first is database products. The company is the leading seller of database management systems for microcomputers with a 60 percent market share. The key product in this line now is dBASE III Plus.

The second product line is Framework II. It is an "integrated" software package that combines word processing, spreadsheets, database, graphics, and communications capabilities into one product.

The third product line revolves around Multimate International, which Ashton-Tate acquired in December 1985 for $23 million. Multimate produces a word processing package with a growing market share in the corporate microcomputer word processing environment.

The fourth product line is presentation graphics, which Ashton-Tate acquired by buying Decision Resources. Its four products are: ChartMaster, SignMaster, Diagram, and Map.

These four product lines have enabled Ashton-Tate to grow 156 percent—from $82.3 million in 1984 to $210.6 million in 1986. The addition of Multimate International's sales adds $40 million to revenues. Decision Resources adds another $15 million per year in revenues. Profits showed an even more dramatic increase from $6.5 million in 1984 to $30 million in 1986, over a 360 percent growth rate.

Stages of Growth

In reflecting upon Ashton-Tate's meteoric growth, four stages of growth or eras can be identified as having distinguishing characteristics in the company's short history. These eras are directly related to changing market conditions and requirements for new marketing strategies.

The first era was *Garage Shop*. Like scores of other garage startups in Silicon Valley, Ashton-Tate had humble beginnings.

The next era was *Ruling Prince*. This stage occurred when the founders hired a manager to bring a degree of "professionalism" to the company.

The third era was *Organized Business*. In this stage, the company significantly enhanced its ability to be a viable long-term company.

The fourth state of the company was *Growing Corporation Era*. The company eliminated a great deal of the risk of survival and focused on becoming one of America's big corporations (see table 8–1).

As the company moved through these various stages of growth, its strategies and tactics changed. Because of Ashton-Tate's hypergrowth, many management and marketing problems have been magnified. For example, the task of hiring 410 people during a year's time, infusing them with the Ashton-Tate culture and getting them to be productive fast enough to keep up with an explosive market has been both an exciting and exhausting challenge.

The following eras of growth describe Ashton-Tate's own story. Many of the experiences, requirements for moving through the stages of growth and marketing problems the company faced, however, will characterize many, if not most, high-technology, fast-growth companies.

Garage Shop Era

The two founders of the company, George Tate and Hal Lashlee, established Ashton-Tate principally because they did not want to pay full list price for software. Hence, they founded a company and made arrangements with suppliers to buy software cheaply. Fortunately, they met a programmer, Wayne Ratliff, who had written a database program called Vulcan, the predecessor of the dBASE product.

Getting attention for the product was the critical marketing issue at this stage. There were hundreds and sometimes thousands of companies competing in the new marketplace. Having the best product is certainly impor-

Table 8–1
Stages of Growth at Ashton-Tate

Eras \ Issues	Time Frame	Factors for Market Success	Type of Market	Market Problems	Management	Culture
Garage shop	Founding of company	Getting attention for product	High margins Growth market	Forecasting Market demand	Entrepreneurial founders	Family
Ruling prince	Years 2-3	Installed base Distribution channels	Explosive growth	Second products; supplying the demand; repositioning the company	Professional manager, realignment of culture	Extended family
Organized business	Years 4-5	Customer service, support, training Perception of success	Oligopolistic Slower growth	Meeting customer needs	Team effort; emphasis on communication; organizational structure	Introduction of rules and procedures; performance focused
Growing corporation	Current	Strategic market planning Anticipate product introduction	Mature	Linking needs with technology development	Motivation; more emphasis on research; manage expectations	Innovation within corporate structure

tant. But it is not the key success factor. Creative advertising, receiving positive product reviews and encouraging word-of-mouth in the young software development community contributed significantly to the product's early success.

During this stage, it is difficult to determine whether a company's success is based on the founders' vision and hard work, the product's capabilities, the need in and growth of the marketplace or a combination of all of these. At this stage of development, Ashton-Tate and many competitors survived despite themselves because of high margins in a high-growth market.

Product success in the marketplace can cover a multitude of mistakes. With a home-run product like dBASE, financial controls are not overly important because money just keeps rolling in. In fact, it is even difficult to spend the money fast enough to become unprofitable.

A difficult task at this stage is to forecast market demand. In a fast-growing market, every forecasted demand is exceeded. It's easy for a company to become self-congratulatory because the pent-up market pull keeps the product exceeding the forecast. Success depends on keeping up with demand and sales dictate company growth.

Starting the Family

Most start-ups quickly develop a "family culture." Ashton-Tate certainly did. Stock ownership, stock options, and other benefits are principal ways to attract and retain the commitment of the first hires. Indeed, these original employees expected to be treated like family.

As Ashton-Tate continued to grow, however, the family treatment approach caused problems. In later stages, as the company hired more employees and sales continued to grow, each individual's scope of contribution to the company inevitably began to contract to some degree. In a ten-person company, employees know everything going on with everyone else. They can see the direct result of their efforts. This is impossible as a company becomes large.

Leadership at this stage is supplied by the founders. Respect is earned by his or her raw energy. Many entrepreneurs succeed when hundreds of intelligent people tell them they will fail just from the sheer raw energy of the entrepreneur, who is convinced he can succeed.

Innovation was the spark for Ashton-Tate in the Garage Shop Era. The product was a bright idea, sales exploded, and success was achieved. Very little attention was given to the future of the product line or to what the company intended on becoming. The innovators were the founders; they fulfilled the roles of president, managers, and marketers simultaneously. Given the pressures just to keep up with the market, they tended to give little attention to building a company.

Ruling Prince Era

This stage for Ashton-Tate was characterized by explosive growth. The company was able to maintain product leadership as subsequent releases of dBASE kept pace with technology and the market. The company's installed base and distribution channels became key success factors. The development of brand loyalty combined with the latest technology and the company's own momentum brought new buyers to Ashton-Tate.

New product opportunities abounded as the market exploded. Supply began to catch up with demand. Given the cash flow, the company had the ability to experiment and produced many marketplace failures. Some of the second products, such as "Bottom-Line Strategist," "Friday!" and "Financial Planning Language" are programs customers will never likely hear about and never find in any store.

The original entrepreneurs needed and could afford to bring in what they hoped would be professional management. At the time, professional management was viewed as the savior, the business Messiah, who would make order out of the chaos.

In many cases, the introduction of a professional manager in a fast-track company comes about because of the demands of venture capitalists or because of his or her credentials or because he or she "hits it off" with the founders.

Becomes Lord and Master

In Ashton-Tate's case, the professional manager took on the powers of a ruling prince. If one disagreed with the ruling prince, he or she was out. If one made a mistake, he or she could be out. Although the company needed professional management, there was little attention at this stage to procedures, budgets, or the development of management depth.

During this stage, management development is the key to making a company a long-term player. In the Garage Shop Era, the company can run on raw power and market momentum. But at the Ruling Prince stage, the opportunity develops to build the management team and put in the financial, managerial, and marketing mechanisms that will enable the company to succeed in the future. If these are not implemented, then the chances of continued growth are slim.

In the Ruling Prince stage in Ashton-Tate's growth, the founders still continued to be the folk heroes. George Tate moved up to chairman, epitomized the culture, and maintained the initial spirit of the company.

The professional manager had the responsibility of looking after day-to-day operations of the firm.

As the number of employees grew, the culture began to become that of the extended family. Some planning systems developed but only around major events. As a result, the company's next product was developed with little attention to announcement procedures or planning that might help future efforts.

The marketing thrust at this stage was an attempt to replicate the success of the first product. In the software business, many companies have had difficulty trying to produce a second successful product. Lotus 1-2-3 and dBASE II launched companies to stardom, but duplicating the success of the first can be, and usually is, extraordinarily difficult.

The difference between being part of the founding team and entering the company at this stage is remarkable. In a start-up company, one has higher highs or lower lows—and they happen with greater frequency. Also in a start-up company, it is possible to build the culture to reflect the founder's perception of what the company should be. If that proves to be wrong, it is relatively easy to change the culture. But at a later stage of growth it is more difficult to alter the corporate culture.

When Ed Esber joined Ashton-Tate, the Ruling Prince culture was far different from what he wanted it to be. His eventual replacement of the Ruling Prince posed an unusual transition for the company. Rarely does a company encounter a succession in management where the successor is so different from his predecessor.

Thus, one of the greatest challenges at Ashton-Tate during this stage was to realign the culture and reposition the company for long-term success. This meant trying to implement efficient and effective management and marketing processes while simultaneously allowing for creativity, innovation, and risk-taking.

Organized Business Era

During this stage, instead of encountering explosive growth, Ashton-Tate experienced merely rapid growth. From 1983 to 1984, the company grew 100 percent; from 1984 to 1985 it only grew 50 percent. In high technology businesses, such growth can be normal for a successful company.

By the start of the Organized Business Era at Ashton-Tate, the shakeout in the software industry was complete. The industry went from hundreds of companies (such as Visicorp, which dropped from being a $40 million star to bankruptcy in two years) to what is presently an oligopoly. The three leading players in the business are Ashton-Tate, Microsoft, and Lotus. These companies vie with each other to maintain leadership position. Their approach is more of an attempt to edge out each other in niche markets than knock off the leader.

In the Organized Business stage of growth, product leadership is less important than the ability to surround products with services that customers need. Many people talk about being a technology leader or a marketing leader. Management at Ashton-Tate believes that a company in today's competitive environment cannot be one without also being the other. It is critical to be both a technological leader and a marketing leader. The big question is, who leads internally?

Give Customers What They Want

At Ashton-Tate, the decision is not to have strong marketing and weak technology or vice versa. Rather, the corporate objective is to lead. Because this is a company that bases its technology on what the marketplace wants, marketing came to lead the company.

Remember, up to this stage, Ashton-Tate did not get customer input. It just announced a product. If it happened to sell, great! Initial product decisions were based on luck, vision, and insight.

However, as the company grew, as competition became keen, and as some products failed, the requirements for market leadership changed. Maintaining market leadership came to increasingly depend on getting input from customers and then meeting those customer needs.

This change corresponded with a shift in the microcomputer software business as the focus moved from selling to individuals to selling to corporations.

Corporations tend to be conservative. They want more than hand holding. They expect methodical and long-term support. Consequently, market success depends on surrounding the customer not only with the product but also with service, support, and training.

A Successful Image Is Important

In tapping the corporate market, Ashton-Tate learned that corporations like to buy from successful companies. Consequently, public relations has become a key factor in the software industry in achieving this goal.

When Ed Esber joined Ashton-Tate, he told the company's PR firm that he did not want to become a persona. People like Mitch Kapor and Bill Gates have been the personas of the software business.

Although he did not want to become a public figure, it turned out the company needed it. Customers and investors wanted to know some person they could say was the guiding light for the company. Internally, however, management tried not to lose sight that a team effort makes a company successful.

The corporate culture continued to change in the Organized Business Era. The company tried to keep the family feeling and entrepreneurial spirit of earlier eras, while at the same time prepare for strong competition in the market.

To change a culture, it is necessary to shock it sometimes. For example, when Ed Esber first came to Ashton-Tate, he decided to lay off fifty people. The message was that one gets paid for performance. In a fast-growth company, everyone has to work extremely hard and it is demoralizing to look at the next desk and see somebody who is not carrying his or her weight.

Leadership at this stage was provided by the president and top management team. The marketplace drove the innovation process, which became product focused. The innovators in the organization now tended to be the product managers and product developers. No longer did all the ideas come from key individuals or the management team. In other words, the company was developing an organization capable of moving the business ahead.

Founding Family Feels Stifled

At this stage the original employees began to express their strongest criticism. They were concerned that management was stifling innovation, requiring too much paperwork, and generally getting in the way of their jobs. One original employee who recently left the company characterized Ashton-Tate as "a corpse being dragged around."

Such radical comments represented how the old culture felt about organizational rules and procedures. It was important not to ignore their concerns. Many of these old-school employees are innovators or entrepreneurs the company did not want to lose. They were comfortable working in an environment without management. They did what they pleased. Yet, as the company grew bigger, management and motivation became new and critical components.

As a result of the growth to an organized business, a classic conflict developed in the organization. The firm's programmers, who tended to be independent and disdainful of paperwork, became overpowered by the marketing people. They resisted the marketing people because they were the ones who had created the initial success and thus believed that such success gave them license to write other products.

Balancing procedures and processes to allow them to focus on their work has required a tier of management that insulates most of the programmers from the rest of the bureaucracy.

Growing Corporation Era

Ashton-Tate is moving into its Growing Corporation Era. At this stage in the growth, the marketplace has become less fragmented. It is easier to determine the categories in which to compete and thus bring products to market more rapidly. As product complexities increase and development cycles become extended, barriers to entry into the business increase.

Ashton-Tate's objective is to be a major player in the computer services and software industry. Through the earlier stages of the company's growth, there was plenty of "career growth" for everybody.

Yet being a twenty-year-old vice-president in a ten- to twenty-person private company is considerably different than being one in a thousand-employee public company. Consequently, the company focus is to invest in people and recruit executives with the experience and know-how to continue to move the corporation.

Before 1985, Ashton-Tate had no strategy, no long-term idea of where it was going. It was strictly responding to its own success. In 1985, Ashton-Tate began a five-year plan. At the time many industry observers considered it a futile effort. How can a company plan in an industry where two weeks has always been considered long term?

Looking Ahead

However, the industry has matured. As a long-term player, Ashton-Tate is looking out five years, attempting to anticipate technological and marketplace changes to better determine where it wants to be.

Innovation will be a key to the firm's success. Remaining competitive will depend on its ability to motivate product managers and developers to be innovative. In the past, companies in the software industry have done very little research and have focused mainly on development. Ashton-Tate intends to place more emphasis on the research side.

Finally, at this stage, the company is beginning to manage expectations. If the company actually continued to grow at the rate it has in the past, it would begin to exceed the gross national product of the United States within a few years!

Unfortunately, many people in growing companies tend to expect 100 percent to 200 percent growth per year. Such growth just does not happen.

Managers at Ashton-Tate are accustomed to doubling their organization or empires every year; that expectation needs to change. Following technology curves and introducing products at just the right pace and time will be essential. In this business, a company can literally produce a new release of a product every six weeks, every six months, or every year. Such a pace, however, will never allow a company to fully realize the profit in the product.

Linking real market needs with technological growth is the objective. As an industry leader, pacing development with marketplace needs is the key to providing the company with the revenue and growth that can allow Ashton-Tate to flourish in a high-growth industry.

Part III
Marketing Techniques for Technology Companies

9
Marketing Technology-Intensive Products to Industrial Firms: Developing a Service Orientation

Rajendra K. Srivastava

While technology-related trade journals are full of success stories about factory automation, artificial intelligence applications, and technological breakthroughs in fields ranging from geology to electronics to metal fabrications to adhesives, industry surveys indicate that applications of these technologies have been limited. "Most literature deals with state-of-the-art developments, but manufacturers want state-of-the-practice applications," says one mid-size company manager.

Many companies appear unwilling or unable to successfully adopt the advanced systems that promise to enhance their competitive position.

A lack of cash is far from the only barrier to the adoption of advanced technology. In a study conducted by Boston University,[1] 90 percent of automation experts surveyed agreed that "there are technical challenges in the achievement of computer integrated manufacturing, but the toughest problems at this time are managerial." About 70 percent of the respondents believed managers were postponing decisions because they were highly *uncertain* about the magnitude of potential benefits.

Fear Leads to Procrastination

Also, as trade journals herald technological advances, the mass media publicize plant closures and manufacturing snags at General Motors' "factory of the future." Regardless of which image is true, the conflicting information strikes fear in managers considering the adoption of technology-intensive products that typically require huge capital outlays.

This perception of risk, when combined with unfamiliarity with new technology, naturally can lead to procrastination or an attitude of "let's wait and let somebody else demonstrate the viability of the new technology."

Experts agree that most advanced technologies are difficult to justify using traditional capital budgeting methods.[1] Artificially high hurdle rates, coupled with the assumption that the company will maintain its current market share and profitability if no changes are made, are part of the problem. Other problems are: the lack of in-house expertise; the absence of internal champions of the new technology within customer firms; a mismatch between promise and performance, and lack of standardization that limits the ability of managers to compare competitive products.

While other factors such as potential cost savings, increased safety, product quality improvement, allure of new technology and desire to stay competitive tend to facilitate the adoption of high-tech products by industrial firms, they are often overwhelmed by the barriers to adoption. This presents a challenge to marketers of technology-intensive products, who must overcome fears before customers will pay serious attention to the advanced technology.[2]

In the following pages we will discuss the factors that facilitate and inhibit market growth. Our premise is that the successful use of new technology requires meeting the needs of and clearing hurdles posed by *all* concerned parties within a customer firm: operational, human resource, hardware/software, and marketing considerations. In doing so, we will draw upon two marketing studies—one in robotics,[3] the other in artificial intelligence.[4] Interestingly, there are common themes across these industries that lead to similar marketing implications.

A Conceptual Marketing Framework

First, any new technology must compete with existing methods of performing the same tasks (see figure 9–1). While the customer is amply familiar with and has used existing technology, his or her perceptions of the new technology may be inaccurate or biased. Nonetheless, the perceptions of the managers represent reality. The (perceived) relative advantages of the new technology over the old, such as productivity gains, increased safety, quality enhancements, serve to facilitate or motivate adoptions (see figure 9–2).

But the perceptions of the capabilities of the new technology are based on imperfect information and, owing to a lack of understanding, are often formed in the absence of a good framework of a set of evaluative criteria. These perceptions, in addition to being inaccurate, also include uncertainty regarding the potential advantages of advanced technology.

Naturally, perceptions of technological capabilities are based on both formal and informal communications. They are, therefore, colored by information presented by the seller, available published information on the capabilities of the new technology, and reports about the actual performance in the field. The uncertainty owing to incomplete understanding is often compounded by the uncertainty associated with incomplete information. Further, promises or representations made by sellers when not matched by

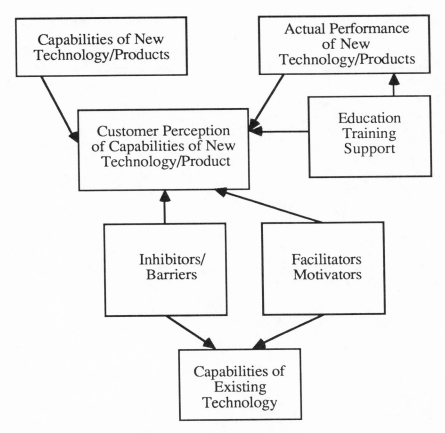

Figure 9–1. Role of Inhibiting and Facilitating Factors

actual technology/product performance enhance this perceived risk. A situation in which costs are relatively well defined and benefits only vaguely defined erects barriers to adoption and, understandably, leads customers to hesitate.

Perceived risk will have several components, including financial (related to cost overruns, time delays, overestimated productivity gains), and political (that is, why should managers stick their necks out to champion unproven technology when rivals stand to gain from their errors in judgment?). All serve to delay adoption. Universally, literature from psychology,[5] sociology/diffusion of innovations,[6] economics/business/finance,[7] and production/engineering[8] illustrate the debilitating role of uncertainty and risk. In short, perceived risk acts as a barrier to adoption. It must be alleviated before manufacturer representations of "enhanced capabilities" over old technology will have serious impact.

Inhibitors/Barriers	Facilitators/Motivators
Operations:	**Operations:**
Compatibility	Improvement in quality
Applications expertise	Increased flexibility
Installation disruptions	
Financial:	**Financial:**
Use of ROI/paycheck	Reduced operating expense
Accounting/cost allocation	Increased productivity
Sticker price shock	
Human Resources:	**Human Resources:**
Lack of skilled personnel	Increased safety
Fear of new technology	
Attitude toward risk	
Hardware/Software:	**Hardware/Software:**
Maintenance	Relative advantage/
Lack of standardization	Existing technology
Product obsolescence	Specific features
Sales/Marketing:	**Sales/Marketing**
Prior "overselling"	Reputation/image
Ability to evaluate competitive	Internal champions
product information	Perception of continued support

Figure 9–2. Inhibitive and Facilitative Factors

Reducing Perceived Risk

Clearly, marketers must play a role in reducing the perceived risk.

First, because advanced technologies are complex, potential clients need to be taught its potential advantages and drawbacks. They also have to be taught to select from among the available competitive options.

Second, the manufacturer must play an active role in the initial installation/design as well as continued maintenance/support of the installed system. This "hand holding" not only reassures nervous customers, but also ensures that high-tech systems are properly used and that customers will find additional uses for similar technology.

Customer satisfaction will result in positive, as opposed to negative, word-of-mouth advertising. Customer education and service/support therefore not only determine the effectiveness of marketing efforts with respect to a specific customer, but because the perceptions of other potential customers are colored by the success or failure of past implementations, they generate additional business. Additionally, education services provide a foot in the door for generating new business. Maintenance and support services provide continuing contact for cross-selling additional equipment and new systems.

This need for a service orientation is further justified by the fact that technology-intensive products are produced (designed, customized as per customer requirements, maintained, programmed, debugged) and consumed or used *simultaneously*.[9] Traditionally, however, production and consumption of goods have been treated as separate, with marketers helping in the exchange process. This attitude is reflected, for example, in the robotics/ automation industry where some manufacturers leave the service function to "value added retailers." In such a situation, customers become dependent on vendors or in the instance of larger firms, on consultants.[10]

More Services Desired

One would expect that as customers gain greater experience with advanced technology, they would want fewer services, but at least in the area of robotics, larger firms and those with more robotics experience are even more interested in service features than other potential buyers. This apparent contradiction may be explainable when we consider that one must experience the need for outside help before one recognizes it. Nonusers, not having encountered situations in which they need external help, may mistakenly feel more confident about their ability to handle problems with in-house technicians.

Because technology intensive products must often be custom-designed to suit the needs of the customer, there is an overlap between the production and consumption functions. Marketers must therefore coordinate the activities among several kinds of parties, including technical staff, users, and managerial personnel representing both the product manufacturer and the customer. In the services industry, there is a growing recognition of the need for "internal marketing," the logic being that personnel representing the seller must believe in (have to be convinced about) the advantages of new products before they can effectively communicate the advantages to customers. Additionally, manufacturer personnel, particularly nonmarketing staff, should be sensitized to client needs.[9]

The next two sections examine how education and support services can help in overcoming barriers to adoption and in enhancing motivation for adoption.

The Nature of Industrial Markets

The markets for technology intensive products such as robotics and artificial intelligence are relatively young and highly competitive. Applications are still evolving. Market growth can be explosive, but it is subject to radical changes because of reliance on the capital budgets of customers who must operate in increasingly uncertain economic environments.

Unlike other traditional industries that first developed products to meet the needs of a "mass" (unsegmented) market, both the robotics[3] and artificial intelligence[4] industries are characterized by diverse user needs and are segmented in the early stage of the product life cycle. Other characteristics that describe these industries include:

- technological uncertainty
- strategic uncertainty
- relative newness of products
- high initial cost, with significant cost reductions with time and number of units produced.
- high inventory costs
- low-volume, high-price products
- embryonic and spinoff companies
- lack of established industry standards
- short time horizons
- products with both hardware and software components
- a high percent of first-time buyers

One must be careful in defining the industry itself. For example, within robotics, some segments such as heavy-duty welding robots are quickly maturing, while the segment for sophisticated robots used for electronic circuit assembly are still in their infancy. Similarly, within artificial intelligence, the nature of the markets for expert systems depends on the area of applications area (such as medicine or geology) as well as the type of end-user (such as academic, government contractors, private industry). Finally, there is a dependence between different industries. For example, the robotics industry has benefited from advances in vision systems emerging from the artificial intelligence industry.

The rapid technological advances in these industries makes products obsolete even before they have had a chance to prove themselves in the field. This represents risk not only to the manufacturers of these products but also to the users, who seem to prefer to wait for better products and lower prices. Additionally, the implementation of these technologies at the user end requires skilled personnel who have intimate knowledge not only of the application area, but also of the new technology itself. This, as discussed earlier, poses a marketing challenge—suppliers must educate the users, either directly or through vendors.

While the promise of technology intensive products is alluring, these products represent significant costs to both developers and users and, given the uncertainties, real risks. The next section examines these factors in depth.

Operations

Operational factors inhibiting growth are primarily related to the application/implementation of the new technologies. They include compatibility of the new technology with established practices or experience; lack of applications expertise within the customers organization; and installation disruptions.

Compatibility of the new technology with customer experience represents a major concern. For example, in expert system applications, computer systems most available to users were DEC or VAX (66 percent), followed by Symbolics (21 percent) and Xerox (5 percent).

Not surprisingly, expert systems with software based on DEC such as OPS-5, GLisp, and Emycin had higher than average market share in the segment familiar with DEC systems. Users familiar with Symbolics gravitated toward KEE. Finally, users not familiar with "AI"-oriented computer systems were attempting to develop their own PASCAL-based systems, for example, to deal with emerging problems.

This tendency to rely on past experience is partly a function of the fact that most available expert system shells are not adequately supported. This is not suprising because software development has primarily occurred in academic institutions, which are in the knowledge—not applications—business.

Lack of applications expertise is primarily software related. For example, 34 percent of the respondents to the robotics survey expressed the need to use consultants to help with applications. This is often coupled with lack of easy-to-use software. While users have gravitated toward consultants for programming services, they would prefer to do their own programming. Fully 21 percent of respondents (and 43 percent of nonusers) gave lack of in-house expertise as a major reason for not using robotics. Similarly, shortage of knowledge engineers/lack of in-house expertise was the second most important factor inhibiting the use of expert systems.

Installation disruptions represent another major risk to users. This is related to disruption of production while the new technology is being installed, as well as installation delays and cost overruns. Clearly, it is much easier to justify robotics installations when the existing operation is manual than it is when it is semiautomated. While it represents a major problem in the robotics industry, it is less of a concern in AI/Expert Systems where the new technology can run parallel and where more time is generally available for installation and testing.

Initial motivations for adopting new technology are generally related to gains in productivity or cost reductions. However, users are increasingly realizing that the byproducts of new technology are equally important.

For example, 81 percent of users reported increased safety of operations. Robots are now used successfully for "3-D" work—dirty, dangerous, and demeaning—thus tending to reduce injuries and exposure to toxic/hazardous materials.

Smaller, but significant, proportions recognize improvements in quality and enhanced flexibility as motivating factors. The recognition of flexibility is an important step because it makes it easier to justify the allocation of resources to expensive technological products.

Financial

Because they perceive high-tech items as high risk, many potential users rely on simple payback and return-on-investment criteria. These criteria have a short-term focus and do not recognize the qualitative (or at least hard-to-quantify) advantages these products deliver. For example, one robotics industry expert noted that accountants have treated robots as if they were hard-dedicated automation projects. How does one account for or allocate costs, when the robotics system is working on something different every day, week, or month? Further, learning curve effects are not considered.

It may not be appropriate to use the same accounting methods for high-tech operations that are now used for standard manufacturing operations (especially if the standard operations are labor intensive). For example, labor costs tend to be paid out of operating budgets while a substantial proportion of automation costs have to be paid up front out of capital budgets.

Also, the ratios of fixed costs to direct labor or overhead costs to direct labor costs change. Managers who are evaluated on the basis of these relationships may, therefore, resist the introduction of new technology.

"Sticker-price shock" occurs when the full cost of the robotics system (including hardware, software, and support services) become apparent to the potential user. The total system price may run two or three times the price of the robot alone. There is often a tendency on the part of managers to allocate all these costs to the capital budget. But, in reality, the cost of support services is a variable and easily can be transferred to the operating budget. Further, increased productivity reduces the operating expenses and it may be argued that part of the costs of operating automated systems should be paid from the operating budget. Because operating budgets are more controllable by individual managers, a strong case can be made for encouraging the separation of the systems price into upfront/capital and recurring/operating components.

High-tech products such as robots and expert systems typically reduce expenses and increase productivity. Our survey suggests that, given the desire to stay competitive, motivation to adopt will grow if suppliers can tell prospective adopters how they can achieve productivity gains. For example, 40 percent of respondents to the robotics survey (and 74 percent of adopters) cited the necessity to stay competitive as a major reason for considering the adoption of robots.

Experience curve effects have shown that robot prices and costs are declining with time. This is primarily because of the integration of research ideas into production, more efficiencies in production, and the reduced costs over time for the high-tech component parts. Equally important are the increases in operating efficiencies at the user end as users learn more about hardware and software and become more proficient in developing their programming/applications software.

Additionally, there is a growing recognition that in the high-tech context standard, investment-decision criteria such as return on investment are too simplistic and inadequate. Managers are beginning to evaluate projects with qualitative criteria that are hard to quantify (such as worker satisfaction, quality improvement, increased safety in the case of robotics, and retention and replication of expertise in the context of expert systems). Unfortunately, these qualitative criteria are recognized and used by only a small proportion of firms. Because most of these are experience-based factors, they are more likely to be recognized and used by firms that have already adopted the product (potential second-generation buyers) and not by those considering high-tech products for the first time.

Admittedly, aspects such as quality improvements are hard to quantify in terms of cost savings or sales increases, particularly the latter. Improvements in safety records are established after implementation and therefore result in insurance savings down the line. Clearly, the high-tech supplier needs to educate the potential buyer about the need to use both qualitative and quantitative criteria. It would be in the supplier's interest to assemble all pertinent information to help "quantify" qualitative criteria.

Human Resources

Lack of skilled robotics engineers/designers, systems engineers, programmers, operators, and technicians (and knowledge engineers in the expert system context) inhibits growth because companies will not invest in new technologies if they cannot adequately use them. Lack of in-house expertise is one of the main reasons robotics/AI consultants have proliferated. Even

General Motors, with the largest and most experienced in-house robotics staff, relies on external consultants.

Fear, or lack of understanding, of technology is a pervasive attitude in many of the older industries that resist change. This is compounded by perception on the part of managers that they will lose control to technicians. Additionally, there may be resistance on the part of organized labor—again in older industries.

The short-term performance emphasis in U.S. industries stacks the deck against innovative technologies. Risk-averse managers typically want to wait until they are more certain that investments will pay off.

The only "human resource" factor fostering market growth is the presence of "champions" of that technology within the customer firm. It is advisable to identify and to cultivate these individuals by furnishing them with information of "ammunition" to wage an internal battle.

Hardware/Software (Product Offering)

Maintenance of advanced technology systems is perceived to be a problem, not only because skilled personnel—in short supply—are required, but also because downtime of these systems is expensive. Only 50 percent of respondents to the robotics survey believed that in-house maintenance of robots was feasible. As expected, this percent was even higher for nonusers.

In the area of expert systems, the large majority of expert system language bases emerge from academia and tend not to be supported. Debugging and system maintenance is therefore a continuing problem and the major source of frustration for users.

Table 9–1 presents the relative importance of support services in robotics and AI markets. The table demonstrates the importance of three components of services that should accompany advanced technology systems/products: installation/design consultation, maintenance (debugging, repairs, hotlines), and training.

Lack of standardization can be confusing to first-time buyers who may hesitate to invest in a system when they feel unable to evaluate available options. This is compounded, in the fragmented high-tech industries, by the fear that the supplier may not be committed to the market and may terminate operations in the future.

In such an environment, it is desirable for the supplier to become "the industry standard" since the entry of large firms (such as IBM, in the case of personal computers) are an important threat.

It is also important to recognize that as industry standards evolve, it does not pay to be different—a lesson well learned by Apple in the microcomputer industry and by IBM in the minicomputer market.

The evolution of standards is especially important in the expert system/

Table 9–1
Relative Importance of Support Services

Services:	Robotics Percent ranked		Artificial Intelligence Percent ranked	
	1	1-3	1	1-3
Installation/Design/Vendor consultation	41	72	38	86
Maintenance with 48 hours/ Debugging/Repairs	23	79	11	49
24-Hour hotline	6	55	9	51
Programming help/User workshops	18	55	3	11
Training	8	51	16	73
Financing	2	13	0	2
Newsletters, Applications news	2	4	1	4
Program updates	N/A	N/A	22	77

AI field, where the transportability of software across hardware is a major problem. Users are unwilling to write off their past experience (since learning new systems involves "unlearning" existing ones).

It is important to recognize that different market segments will have different needs and therefore will have different emphasis in evaluating product features. For example, in evaluating robotics, sensing systems (vision, tactile) and repeatability are far more important in the electronics industry than to other industries. Similarly, natural language capability and display graphics are more important to industry users than to academic ones.

Sales/Marketing

Previous marketing efforts have not always been successful or well received by industry because technology intensive products have been oversold as the panacea of productivity ills. Similarly, marketing efforts often do not deal well with problems related to first-time installations—for start-up problems that cannot usually be foreseen by users.

Customers want high-tech products such as robots and expert systems to be "proven." But, in competitive industries, it is hard—for proprietary reasons—to arrange demonstrations of successful installations. Providing "loaners" may help but could be counterproductive unless accompanied by support services. Both (loaners and services) are expensive and represent substantial selling costs that can be justified only for large projects.

The reputation and image of the supplier, together with the potential adopter's perception of continued support by the supplier, are prime factors for facilitating adoption. So, too, is the continual cultivation of internal support. The marketing process is usually lengthy, requiring external and internal efforts to provide information to all parties involved in the decision process.

Marketing Implications

General marketing strategies and actions are summarized in figure 9–3. Specific recommendations include, for example:

- encouraging customers to assemble a project team;
- stressing supplier support by pointing out that the new technology/product will be accompanied by programming, training, installation, design, and maintenance services;
- considering avenues for demonstration and trial;
- encouraging customers to allocate part of the systems cost (especially services) to the operating budget.

These strategies require serious effort on the part of suppliers to "educate" customers about the use of technology intensive products; to provide training and help during installation and design of (often customized) systems; and to ensure continued maintenance/support.

After-sales support is important not only for getting new orders from existing clients, but also for developing a reputation that helps generate new customers. Applications engineering and missionary sales are highly important in the early stages of adoption of most new technologies. Naturally, provision of support services and design/installation of systems requires the supplier to understand the customer's situation so as to develop appropriate systems.

Deal with "Perceived" Risks

In designing marketing programs, suppliers must attempt to address "perceived" risk first if they are to fully "sell" the advantages offered by the new

Operations and Human Resources

• First installations should be new additions rather than replacements of existing technology. Implement initially in environments where productivity is lagging and where gains would be obvious. Allow plenty of time for debugging.

• Encourage clients to assemble a project team that interacts with the supplier's technical staff to enhance learning and to establish internal support. Concentrate on ongoing adaptive learning to ensure that the customer gets full use of technological capabilities.

• Stress product functions and attendant service packages include programming, training, installation, design, and maintenance.

• Consider avenues for demonstration and trial, including loaners accompanied by services (a leasing arrangement for a limited period may be viable). Show videotapes to demonstrate performance. In training sessions, show mistakes as well as successful applications.

Financial

• Develop promotional literature that shows how to balance both qualitative and quantitative considerations. Do the customer's homework.

• Justify the cost of the new technology by taking into account both the relative advantages and risks--concentrate on value-based pricing.

• Encourage clients to allocate costs partly to the operating budget--productivity gains decrease the operating budget requirements.

Hardware/Software/Product Offering

• Clearly demonstrate technological advantages and ease of use. Stress specific product features that are of interest to the customer/segment.

Sales/Marketing

• Do not oversell. Maximize control over applications. Address perceived risks first. Stress the long-term commitment of the supplier.

• Consider both the type of buyer/segment and the type of applications needed in developing products and services.

Figure 9–3. Marketing Implications

technology (compared to the existing one). In doing so, marketers must coordinate interaction between the user's and supplier's technical and managerial staffs. Particular emphasis must be placed on generating internal support or "champion(s)" who must address the concerns of parties with whom marketers may not often come in contact.

Suppliers of technology intensive products must recognize the interdependencies between hardware and software, which must be compatible with each other, but also (if possible) with the customer's previous experience. This requires a supplier to spend some time learning the needs, experiences, and preferences of potential clients before making formal proposals. The needs of market segments are likely to vary; therefore, customization will be required for hardware/software/service elements of the "product offering."

Because purchase justification continues to be a problem, suppliers must educate the customer in the qualitative factors that enhance the desirability of the products and provide the necessary information to enable the customer to quantify these factors to some degree. It is important that the customer recognize the full costs of the system (hidden costs are frustrating and lead to ill will) and that these be allocated appropriately across both capital and operating budgets.

To enhance the acceptability of systems, it is advisable for the supplier to amortize investment recovery costs over several years by spreading them across both service contracts and initial purchase prices. This will minimize the sticker shock.

Concentrate on "System" Sales

Suppliers of technology intensive products should concentrate on "system" sales (hardware/software/services) for two main reasons.

Leaving some of these functions to other suppliers or consultants results in the lack of control and therefore the potential for negative client experiences—which ultimately reflect badly upon the supplier.

They represent a relatively easy-to-tap source of additional revenues (assuming, for example, that the "product" or hardware is already sold; it is harder to sell hardware to a new client than to cross-sell services to existing clients).

One must recognize that the relevant base for comparison in fragmented industrial markets for technology intensive products is the existing technology, because information availability often does not allow the direct comparison of alternatives. In such a context, the pricing of product offerings should be based on the "value" of the technology to the customer. This will be a function of the relative advantages of the new technology and the additional risk it poses.

Clearly, marketing technology intensive products to industrial firms requires the supplier to be service oriented. The supplier must market the entire system—hardware/software/services—that together leads to successful implementations. Each element is important. If the supplier is not in a position to deliver appropriate services directly, then adequate incentives must be provided to the distribution system (for example, value-added retailers or system houses) to do so.

Notes

1. Huber, R., "Justification: Barrier to Competitive Manufacturing," *Production,* November 1984, 55.

2. Abramson, D., "Robots: Overcoming Fear First," *Marketing Communications,* Vol. 8, No. 10, October 1983, 36–40.

3. Srivastava, R.K., "Market Analysis of the Robotics Industry," unpublished report, Department of Marketing, University of Texas at Austin, 1985.

4. Srivastava, R.K., "Market Analysis For System Builders/Expert System Building Tools," unpublished report, Department of Marketing, University of Texas at Austin, August 1985.

5. Kahneman, D., Slovic, P., and Tversky, A., *Judgment Under Uncertainty: Heuristics and Biases,* Cambridge University Press, New York, 1982.

6. Rogers, E.M., *Diffusion of Innovations,* The Free Press, Inc., 1962.

7. Kaplan, R., "Must CIM be Justified by Faith Alone?" *Harvard Business Review,* March-April 1986, 87–88.

8. Aronson, R.B., "Scarcity of Capital Puts Robot Revolution on Hold," *Machine Design,* Vol. 55, No. 24, October 1983, 66–71.

9. Parasuraman, A., Zeithaml, V., and Berry, L., "A Conceptual Model of Service Quality and Its Implications for Future Research," *Journal of Marketing,* 41, Fall 1985, 41–50.

10. Kozikowski, W., *U.S. Manufacturers' Five-Year Industrial Automation Plans for Automation,* Machinery and Plant Communication Systems, National Electrical Manufacturers Association, Washington, D.C., 1985.

11. Robot User Buying Patterns: A National Market Survey, *Robotic Industries Association,* Dearborn, Mich., 1985, 9.

10
Public Relations Techniques for High-Technology Start-Ups

Dennis Lewis

High-technology start-up companies often approach public relations with unreasonable expectations. Seeing the coverage that leading firms get in the business, consumer, and industry press, these companies assume that the "excellence" of their own products should somehow guarantee them a large slice of the publicity pie. Yet few of them have thought seriously about how much time, effort, and planning goes into a successful public relations campaign.

Creating Credibility

High-technology public relations for start-ups is primarily an educational or communications process directed toward creating credibility for a company and its products in the minds of a broad range of audiences, from potential customers, distributors, and dealers to potential employees and investors.

Though it is true that many ingredients go into establishing this credibility—the depth of a company's management, the company's financial strength, its distribution or sales channels, and so on—the key at the beginning is clear product differentiation or positioning. Young companies that cannot clearly communicate how their products or services are different from or better than their competition's—and how their long-term marketing strategies will keep them that way amid fast-changing market conditions—will never get the interest of the key industry influencers: the editors, analysts, and journalists who stand guard at the gates of public perception.

Only a firm that has achieved clear and positive product differentiation can move on effectively to the next steps in its public relations campaign: credibility based on overall market position and corporate strength. Obviously, credibility can be created in these areas most effectively when a firm has correctly positioned its products and has the satisfied customers and the profits to prove it.

Positioning Starts with Product Design

Many young companies think that product positioning begins only after a product has been developed. In reality, positioning a high-tech product is a dynamic, ongoing process that should begin with its very design. For whom is the product created? What features and benefits must it have to ensure that it will compete successfully against its competition? Do these features and benefits give the company a "communications" edge? And what about intangibles such as aesthetics, ergonomics, reliability, ease of use, compatibility, upgradability, and so on?

The company also must squarely face the issues of documentation, distribution, service, and support. A product's image in the mind of users and potential users, its "persona," so to speak, is greatly colored by how effective a company is in these areas.

In short, a product or service is far more than its generic definition or description—far more than the basic function it fulfills. It takes on its meaning and value in the minds of potential customers based on a broad range of market variables, including wants, needs, expectations, and even fears. To be successful, a public relations campaign must thus take into account the total marketing environment.

Assuming that a new company has done its marketing homework and that the product, including all its intangibles, can be differentiated in a positive way from its competition, it is important to remember that it takes time to successfully portray the actual strengths of a company and its products. Companies that expect instant results are forgetting that the real point of their public relations efforts is to reach and educate the industry influencers, who in turn will influence the marketplace through written materials and word-of-mouth referrals.

Though good product PR can start generating inquiries and sales leads in as little time as three to six months, the real payoff—the proper positioning of the company and products in the minds of potential customers and investors—takes much longer. Companies that overlook this educational, long-term aspect of public relations will never create the credibility that long-term success requires. This is especially true for new firms, which must build their product and corporate credibility from the ground up.

Working with the Press

Establishing positive relationships with the trade, business, and financial press is essential to the long-term success of a new company's PR efforts. Those who treat the press professionally will eventually reap the results for which they are looking: increased coverage of their company and products.

But professionalism requires more than the right attitude; it also requires a thorough knowledge of the way business and industry editors like to work.

Young companies that make their executives readily available to the press, for instance, will find their name in print more often than those that do not. The most successful companies, from a PR standpoint, are those whose executives have established long-term, professional relationships with the editors of the publications they are trying to reach.

These executives take an active role in their companies' PR programs. They foster an open line of communication between the editors' offices and their own. They make every effort to share their ideas on the industry and their reactions to relevant news events with the writers and editors of the key publications. They are receptive to the valuable feedback they can get from leading editors and journalists regarding their products and marketing strategies.

In addition to telephone interviews and on-site meetings, three of the most effective ways of meeting the press are press tours, trade shows, and press conferences. A week-long press tour to the offices of key industry and business editors and analysts in conjunction with a major company announcement can help a new company not only spread the word on its products and services, but also establish important editorial relationships. Likewise, personal meetings in booths or hospitality suites at trade shows offer a cost-effective way to accomplish the same goals. Press conferences are far less personal and are generally not advisable for young companies unless the product to be announced represents a technological breakthrough.

Plan Carefully

Whatever form they take, however, meetings with the press require careful planning. Frequently, new companies assume too much technical knowledge on the part of the press. Many reporters and editors cover broad areas of the industry. As a result, their expertise in any single area is often quite limited.

In planning your presentation to publications, make sure the level of detail is appropriate and that the presentation is flexible enough to take into account the varying interests of the editors you will meet. Discussions with technical publications will be highly product- and technology-oriented, but discussions with business and general-interest publications will tend to be at a much higher level—frequently involving industry overviews and trends.

In any case, be sure your presentation explains your product's or service's function, market, benefits, competition and so on in a clear, step-by-step manner. If your product is a major one or is highly complex, you might consider developing a short slide or video presentation to demonstrate the technology and its impact on the industry.

It is crucial when dealing with the press to realize that ultimately it is the editor and no one else who decides what gets printed. An editor can seldom give you a guarantee of coverage. Placing undue pressure on an editor can backfire. One of the quickest ways possible to get yourself in trouble with an editor or journalist, for example, is to mention all the advertising dollars you are spending in that publication.

"Selling" an editor on a story, however, is not unlike selling a new product to a potential buyer. Editors need news and articles to serve their readers. As with any selling, it is important to remember to "close" the deal. But to do this you need to have a good understanding of the publication with which you are dealing. What are its real needs? How does your product or company news fit in? Whatever you do, don't try to force an "artificial" or unwanted story on an editor.

If the publication is extremely important to you, you may wish to give it an exclusive or a head start on a particular story. By doing so, you may be able to get a more comprehensive story or even score a cover or front page if your product is a major one. But be careful in giving such "head starts." As a new company, it may be more helpful in building important relationships to treat all your key publications in an even-handed way.

In the interview itself, when the reporter's or editor's pen stops it's usually because you've lost him or her, or because the level of detail or direction is not appropriate to the publication. If this happens, don't be afraid to stop and find out what the editor really wants to know.

Under no circumstances assume that putting a press release into an editor's hands will somehow save the day. A personal meeting gives the editor an opportunity to dig out an angle on your "news" that you may not have even been aware of. And it gives you an opportunity to build a productive relationship that will ultimately result in word-of-mouth referrals and publicity.

The Art of Writing a Good Press Release

One of the basic tools used in communicating with the media is the ubiquitous press release. Though thousands of these releases are sent out weekly to editors, journalists, and research and financial analysts, only a few of them ever see the light of day.

An effective press release about an important new high-tech product or some company news such as new funding, contracts, changes in management, and other developments is as much the result of art as it is of journalistic skill. It not only demands clear, well-structured composition—with a telling headline and lead—but it also requires a fresh way of conveying the "news."

A press release must express its message simply and clearly, in language appropriate to the readers the company is attempting to reach. In this regard, releases should be "tailored" for the types of publications to which they are being sent. There is nothing worse than a jargon-filled product release going to an editor of a consumer or business publication or a release without any technical substance going to the editor of an engineering or OEM publication.

Companies, especially start-ups, often approach press releases on high-tech products as little more than glorified "spec sheets" adorned with self-serving quotations from company executives. Even if such releases do somehow see print, they often will have limited impact on their intended audience.

With editors besieged by an ever-growing number of press releases, it is extremely important to make your release stand out in relation to the intelligence of the writing, the accuracy of the news, and the quality of the photographs and other visuals included with it. For new product releases, it is important to say quickly for whom the new product or service is intended, what capabilities it offers, when it will be available, and how much it will cost.

Where "art" can enter the picture is when the writer goes a step further and actually attempts to "position" the product or service in the reader's mind. But this takes knowledge. A writer of a product release who thoroughly understands the industry to which it pertains will usually write a far more readable and informative release than someone who does not. He or she will bring to bear a perspective that will help both the editor and the reader see the product or service in a new light.

Don't forget, however, that editors, like the rest of us, don't enjoy spending their time reading dry, boring material. They want to read interesting material that relates to their own specific publications. But don't try to make your release "interesting" by means of mechanical tricks, such as multiple type styles, colors, and so on.

Make it interesting through the quality of your writing and thought. Above all, convey factual information. A release full of unwarranted "hype" will not only end up in the wastebasket, but it also will have a negative effect on your company's reputation.

Develop a Professional Press Kit

A young company attempting to create credibility for itself among the industry influencers should develop a well-designed, professional press kit that is attractive to the eye.

Such kits typically consist of a folder with dual pockets inside and the company logo on the outside. The folder should be strong enough to hold a variety of materials, including current releases, good-quality black-and-white product photos, price lists, spec sheet, an in-depth corporate backgrounder, a product backgrounder where appropriate, and a fact sheet.

Corporate Backgrounder and Fact Sheet

The corporate backgrounder is generally a five- to ten-page document that gives a comprehensive overview of a company, its history, its products, and its marketing strategy. Because it is designed for editors and is not a marketing piece, it should be written in a straightforward, noncommercial way.

For those editors who do not like to read lengthy backgrounders, it is useful to include a one-page fact sheet on the company. These are generally typeset and include basic information about the company, its products, distribution channels, key personnel, and so forth.

When to Use Press Kits

Press kits are generally given to editors who want an in-depth look at a company. They are particularly useful during trade shows (both in personal meetings and in the press room), press tours, and news conferences.

In putting together an attractive press kit for an event, be sure to pull out old or irrelevant news releases, photos, brochures, price lists, etc. Editors resent having to weed through a lot of old or unnecessary information.

Using Application Stories for Product Positioning

Once a new company has brought its product to market and has actual users, one of the most powerful techniques for positioning it in the minds of potential users is the "application" article or case history. Written for the news pages of technology, trade, and even business publications, these "success" stories of actual users can frequently build credibility to a degree unparalleled by other methods.

Application stories are simple in principle. They are merely news features that show through text and photographs how a company or organization is using and benefiting from a particular product or service. But often they are difficult to execute in a way that will ensure their publication; especially today when fewer and fewer general industry publications will accept them.

There are five basic requirements for a good application story.

1. It should convey *newsworthy* information.

2. It should be *noncommercial;* in other words not be a "sales" piece.

3. It should contain a clear statement of why the company *needs* the product or service and how it was selected from among competing products.

4. The benefits discussed in the story should be *quantified* if possible and backed up with quotations from actual users. These quotes should not, however, sound like "endorsements." Editors and journalists want to know how a product is actually helping a company; they don't want advertising disguised as news.

5. The story should demonstrate the product in *action*. The reader should feel he or she is actually learning how a particular company solved a particular problem by using a particular product.

When interviewing the user for the story, be sure to get the detailed information that will give the story substance. A good application story should give the reader an in-depth look into how the product has helped the company's day-to-day operations.

Before starting the interview, however, be clear on how you want to "position" your product in relation to industry trends and what features or benefits you want to emphasize. Though what you write will be largely dependent on what the user has to say, you can direct the story to some extent by the kinds and quality of the questions you ask.

If the interview is done artfully, you even may learn something new about how your product is perceived in the user's industry and how it can be better differentiated from its competition. Once your article appears in print, you should reprint it and make it available to your sales people.

Bylined Articles Demonstrate Management and Technical Expertise

Articles written for key publications and bylined by company management offer a young company an excellent way to demonstrate its special expertise and position its management at the forefront of the industry. And once the articles have been published, they can be reprinted and used as collateral literature.

Most industry publications are hungry for well-written articles with fresh ideas. The editors actively seek such articles to keep their publications' subscribers abreast of the latest technologies, standards, management techniques, marketing trends, and so on.

In planning a bylined articles program, there are several different types of articles to consider: viewpoint pieces, technical articles, and industry overview articles.

Viewpoint pieces enable a company to speak out on controversial issues in a very personal way. They are excellent vehicles for companies with executives who have strong, controversial opinions and who wish to position themselves in relation to them.

Technical articles are generally product or technology oriented. As such, they enable a new company to demonstrate its design expertise and technical innovation.

Industry overview or trend articles help show that a company and its management have a comprehensive grasp of some aspect of their industry, thus further enhancing their overall credibility.

The secret of getting all such articles published is to work closely and well in advance with the editors of your target publications. Bounce your articles ideas off them and listen carefully to their reactions. By being responsive to their editorial needs you can greatly increase your chances of getting your articles into your target publications.

Taking Advantage of Trade Shows

Though increasing numbers of well-known manufacturers of high-technology products are questioning the value of exhibiting at the popular industry trade shows, many of the young companies that take part in such shows are losing out on the publicity opportunities that these events afford.

Industry trade shows bring together in one location not only the distributors, dealers, manufacturers, systems houses, end users, and others that are so vital to a company's success, but also the leading industry editors and analysts. These editors and analysts attend such shows for a variety of reasons, but the primary ones are to learn about and report on important product and company developments.

The first step to editorial success at a trade show is proper planning. Be sure to find out which publications are planning show preview issues and what their editorial deadlines are. It is also important to look into the show dailies, the publications that distribute daily issues at the show. By working well in advance with these publications your company may be able to score an important interview or article.

At the show itself, a variety of techniques can be used to gain publicity. One approach, of course, is to hold a press conference. At big shows, however, this should only be done if your product or news represents a major breakthrough. A press conference in most other circumstances, especially for a company that is relatively unknown, probably will not be well attended.

Another approach particularly suitable for smaller, relatively young firms is to set up private meetings with the editors who will be attending the show. This will necessitate some homework to find out who will be at the show, but it provides an excellent opportunity for a company's top management to meet the key editors on a one-to-one basis.

A third approach is to host a special event such as a party, contest, or promotional giveaway. Parties are quite common at the major shows and there is a great deal of competition for editorial attendance. Unless a company is well known or does something highly dramatic or unusual, it is unlikely it will get a suitable payback on the substantial investment that will be required. Contests and giveaways, on the other hand, can sometimes help companies increase their overall presence at a show without a large investment.

A cost-effective alternative to the party is the "hospitality suite." A hospitality suite that offers a variety of food and beverages gives editors an opportunity to get away from the hubbub and hard sell of the show and relax or talk quietly with company management over a drink before heading off on the party circuit. It also gives the company a chance to demonstrate its products to editors away from the constant interruptions of the booth.

One of the best ways to create credibility for your company and its key people at a trade show is to chair or give a paper at a technical session or seminar. Depending on the specific show and session, you may be able to reach a large cross-section of industry influencers and potential customers. To accomplish this, however, means working up to a year in advance with the appropriate show management.

Emerging high-technology companies must establish their presence in the marketplace. Public relations plays an essential role in this process. To be effective, however, a public relations campaign requires planning, knowledge about an industry's infrastructure, and an understanding of when and how to use techniques that build credibility and position products.

11

Determining Whether Telemarketing Can Sell High-Technology Products and Services

Joel Leider

Telephone sales has been a way of life for a multitude of organizations over the years. You are no doubt familiar with them and have received telephone solicitations from your alumni association, the local newspaper, commodity brokers, insurance salesmen, and financial advisors. At your office you have received calls from job applicants, office equipment vendors, advertising agencies, trade publications, and head hunters.

Frequent telephone contacts of this nature often are irksome to busy executives who do not have the time or inclination to respond to spurious solicitations. With this as a total experience base, it is not surprising that most marketing executives in high-technology companies are biased against the possibility of using telemarketing as a potential sales tool. More open-minded executives may have given it a try but relegated the effort to second-class status because of the same preconditioned thinking and did so without keeping the entire marketing effort in perspective. This is a real mistake. Telemarketing of high-technology products can work.

Winchester Systems sells high-technology products costing five thousand dollars to twelve thousand dollars almost exclusively by telephone. The products are specialized disk subsystems and are offered to a specific niche of computer owners who are geographically dispersed. Sales are consummated without ever visiting the customer. Repeat business is common and good customer relationships are established over time. Numerous referrals and voluntary testimonials attest to the success of the endeavor.

What Is Different about High-Technology Products?

High-technology products are frequently typified by leading-edge product sophistication using rapidly evolving technology that results in numerous product improvements. These products often are marketed into small niche markets with high initial profit margins but short market windows. Competitors, while small in number, drive prices and especially margins down quickly. Distribution channels are frequently time consuming and expensive to set up and once established may not be effective because of the continuous training and support required for a large number of people, each of whom sells a small volume of product.

Traditional View

Historically, any product worth selling was worth selling face-to-face, especially high-technology products. After all, aren't high-technology products too complicated to sell any other way? Besides, the rationale continues, there is not a better way to establish good rapport with a prospective customer than with a "free" lunch. While this may be true, economic reality means that this style may be a luxury for many high-technology products.

Can Telemarketing Replace Field Sales?

Telemarketing never will make field sales obsolete. Face-to-face selling is still the most effective means of gathering information and influencing others—the critical functions of the salesperson. However, at today's cost of two hundred dollars per sales call, many executives insisting on only field sales forces may be insisting on overkill and undercoverage. Much of what is accomplished in field sales can be done equally well, or certainly well enough to make the sale, and frequently in a more efficient manner, by telephone. frequently in a more efficient manner, by telephone.

Will It Work for Me?

This of course is the crux of the matter. Any singular approach to the market is suboptimal at best. Deciding to use telemarketing is similar to deciding to use advertising or direct mail. One must evaluate the impact on the whole of the marketing effort and integrate the new sales tool into the status quo. Potentially hundreds of factors may affect the decision to use telemarketing as a sales tool. Figure 11–1 depicts many of the more significant and common factors that must be considered. There is no simple formula to make the determination. Ultimately a judgment to try telemarketing must be made based upon these and other factors. A decision to commit to telemarketing is then based upon actual results of pilot telemarketing operations.

Factors	Telemarketing Appropriateness Ranking						

Prospect Demographics

• Total number of prospects	Large	1	2	3	4	5	Small
• Geographic distribution	Dispersed	1	2	3	4	5	Concentrated
• End user of OEM	End User	1	2	3	4	5	OEM
• Account potential	Small	1	2	3	4	5	Large
• Purchasing authority	Low	1	2	3	4	5	High
• Number decision influencers	Few	1	2	3	4	5	Many
• Prospect sophistication	High	1	2	3	4	5	Low

Product Characteristics

• Price	Low	1	2	3	4	5	High
• Complexity	Low	1	2	3	4	5	High
• Time required to evaluate	Short	1	2	3	4	5	Long
• Ability to prove benefits	Easy	1	2	3	4	5	Difficult
• Type of product	Supplies	1	2	3	4	5	Capital Equipment
• Frequency of requirements	Seldom	1	2	3	4	5	Often

Competition

• Product differentiation	High	1	2	3	4	5	Low
• Price comparison	Lower	1	2	3	4	5	Higher
• Market share	Low	1	2	3	4	5	High
• Product positioning	Defined	1	2	3	4	5	Undefined

Prospect's Organization

• Number of decision influencers	Few	1	2	3	4	5	Many
• Size	Small	1	2	3	4	5	Large
• Approval cycle time	Long	1	2	3	4	5	Short
• Importance of purchase	Low	1	2	3	4	5	High

Marketing Mix

• Advertising	Product	1	2	3	4	5	Image
• Public relations	Extensive	1	2	3	4	5	Limited
• Trade shows	Numerous	1	2	3	4	5	Few
• Direct mail	Extensive	1	2	3	4	5	Limited
• Collateral material	Complete	1	2	3	4	5	Incomplete
• Incoming 800 WATS	Extensive	1	2	3	4	5	None
• Field sales	None	1	2	3	4	5	Total
• Manufacturer's representatives	No	1	2	3	4	5	Yes
• Industrial distributors	No	1	2	3	4	5	Yes

Product Credibility

• Product awareness	Low	1	2	3	4	5	High
• Technological credibility	High	1	2	3	4	5	Low
• Testimonials availability	Yes	1	2	3	4	5	No

Figure 11–1. Factors Affecting the Decision to Telemarket

Factors Affecting the Decision to Telemarket

The decision to telemarket is affected by potentially hundreds of factors. Many of the significant and common factors are listed here with a scale to qualitatively indicate how the factor influences the decision. Each situation is unique and thus each factor will carry different weight in different environments. This figure may be used as a worksheet to see how amenable your high-technology product or service is to the telemarketing approach. On a five-point scale, one indicates high applicability and five indicates low applicability.

Certain factors have relatively obvious relationship to a field sales versus telemarketing decision. Geographical dispersion of prospects is one such factor. Visiting a remote prospect may not be feasible for a thousand-dollar sale but may be justifiable and perhaps essential for a million-dollar sale. All significant factors must be considered simultaneously.

In the case of Winchester Systems, the large geographically dispersed prospect base, end user nature of the product and limited account potential indicated the desirability of telemarketing. Relatively clear product differentiation, ease of qualifying benefits, and sophistication of the prospects more than offset the difficulty of dealing with the long approval cycles of capital equipment purchases.

Decision Makers

One key factor, the number of decision influencers, seems to uniquely distinguish itself from the others and is a singularly crucial factor in the analysis. The number of decision influencers especially complicates the ability to sell via telemarketing. Control over the sales situation may be lost because of the inherent inability to influence many people simultaneously, as is possible in a meeting. Conference calls are useful to reach two and perhaps at most three people. Interestingly, price is usually cited as the key criteria for determining whether a product can be telemarketed. Price alone is not uniquely significant. If there is only a single decision maker, he or she often has the authority to make an immediate commitment. Price is only significant to the extent that it increases the number of decision influencers.

Where to Begin?

As a high-technology marketing executive you control the marketing mix (input) and sales process (application of inputs). You thus can have a significant effect on sales and profits (output). This simplified functional relationship is depicted in figure 11–2. To do an effective analysis, it is crucial to understand the intricate relationship of the existing balance between your current marketing mix and your current sales process, both conceptually and factually. Luckily, most of the information required is already in your customer and prospect files or understood informally by you and your salespeople. If you have a good lead tracking system, you have an additional wealth of information at your fingertips.

A typical marketing mix consists of advertising, press releases, trade shows, field sales force, collateral material, demonstration equipment, and presentation materials. Telemarketing is just a new weapon in this already formidable arsenal. The optimal result is to strike a new balance among the tools. In theory, perfect balance occurs when the next dollar invested in any and all of the input resources produces exactly one dollar of profit. Accomplishing the new balance is not at all straightforward, however.

All sales processes have two things in common: prospects and a decision-making process. All times in the marketing mix are intended to obtain prospects or influence the decision-making process. Telemarketing is no different. The key to success is to understand the intricacies of the decision-making process for your product and take a totally fresh look at how to use a new combination of marketing tools that includes telemarketing to persuade decision makers and significant buying influencers to purchase your product.

Empirical Process

The introduction of telemarketing as a new sales tool into the marketing mix requires planning, intuition, and most importantly, experimentation. Introducing an unproven tool into an existing marketing mix may have unpredictable ramifications. Take heart, though, because it is essentially an interactive process, which, through experimentation, careful scrutiny of results, and reiteration, will converge quickly to success or to the conclusion that it is inappropriate. Be humble. Keep investments in new ideas low but give them a fair chance. Testing will generate the information you lacked *a priori* to make the decision to use telemarketing in the first place.

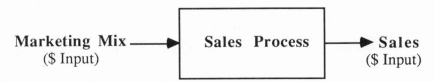

Figure 11–2. Input/Output Functional Relationship

Anatomy of a Sale

A simplified version of a typical field–sales-oriented high-technology sale is shown in figure 11–3. It begins with a lead generated through some form of advertising, promotion, or referrals and ends with a purchase order after some relatively lengthy period of time. Only a small percentage of raw leads turn into sales. It is important to know the attrition rate at each stage so that intermediate results can be monitored when changes are made.

The total sales cost in this example is $14,470 to convert three percent of the initial leads. Winchester Systems sells an average of 1.4 systems per sale at an average selling price of $7,500 or a total expected sales revenue of $31,500 if this approach were used. Projected sales costs using this approach are 46 percent of sales, which is obviously too high. Even counting repeat business of an additional 1.4 unit sales to half the customers within six months, the total sales are $47,250 and sales costs are 30 percent. This approach would be too costly for Winchester Systems products given the current market size, prospect distribution, advertising effectiveness, and prospect conversion rate.

Once you can draw up a chart like this for your products, complete with conversion rates and costs, you have created the basis for an exceptionally insightful analysis. Just preparing this chart will suggest a variety of new alternatives. Now the fun begins.

Sales Activity	Prospects Remaining	Delivery Vehicle	Elapsed Time (weeks)	Activity Cost	Total Cost
1. New lead	100	Mail/Telephone	0	$25	$2,500
2. Literature fulfillment	100	Mail	1	$10	$3,500
3. Qualify prospect	100	Telephone	2	$15	$5,000
4. Initial meeting	30	Field visit	4	$200	$11,000
5. Follow-up call	20	Telephone	5	$10	$11,200
6. Demonstration	10	Field visit	7	$250	$13,700
7. Proposal	5	Overnight service	8	$100	$14,200
8. Additional follow-up	3	Telephone	12	$50	$14,350
9. Approval/Purchase	3	Telephone	16	$10	$14,380
10. Post-sale follow-up	3	Telephone	20	$30	$14,470

Figure 11–3. Field-Sales–Oriented Sales Process

Some changes in the sales process are immediately obvious. For example, one could easily qualify more stringently by telephone, thus reducing the number of initial field visits. Less obvious is the possibility of redesigning the literature to educate the prospect in stages with intervening telephone contacts to requalify and clarify. If a field visit is required, the prospect is very well qualified and more likely to become a customer. Substituting mail and telephone calls for field visits obviously saves time and money but it also dramatically increases the number of prospects a sales person can manage. When the total universe of prospects is relatively large, this is a real advantage.

As long as the effectiveness (conversion rate) of the sales process is not impaired significantly by the substitution, the efficiency ($sales/$marketing cost) will increase by both higher sales and lower costs.

Marketing objectives must be kept foremost in mind before tinkering with the sales process and marketing mix. If the goal is to increase market penetration in a large prospect base, the changes to the sales process will be very different than if the goal is to maximize the conversion rate from a limited number of prospects with high dollar potential.

In high-technology sales, it is a widely held belief that at some point a field visit is necessary to get the sale. It may not be so. It is definitely possible to routinely sell high-technology products without ever seeing the customer. In the case of Winchester Systems, it was considered an absolute necessity to demonstrate the product in the field to make the sale. Retrospective analysis showed that the field demonstration accomplished several crucial aspects in the sales process. Experimentation with alternative techniques yielded an incredible breakthrough—every item on the itinerary for accomplishment in a field demonstration could be achieved in other ways without sacrificing the sale. Winchester Systems totally eliminated the field sales person by carefully defining what is accomplished by the field visit and by making appropriate changes to the sales procedure, collateral materials, and sales policies as outlined in figure 11–4.

Taken collectively, these alternate sales techniques give rise to a different new telemarketing-oriented sales process depicted in figure 11–5.

The resultant sales cost are reduced by 50 percent, to $6,870 from $14,470, with no measurable change in overall sales effectivenes. The new process uses no field visits and makes extensive use of telephone, collateral material, promptly prepared personal letters, and overnight delivery services. Using this approach, Winchester Systems keeps the overall sales cost to under 15 percent, a much more reasonable figure.

It would appear that the overall sales cycle would be significantly delayed by eliminating field visits and substituting additional follow-up calls and literature fulfillments. It does not seem to be so. The sales cycle is predominantly a function of the prospective customer's organization. Large or-

Accomplishment of Field Visit	**Alternative Technique**
Determine need.	Ask critical qualifying questions on telephone.
Create interest in product.	Convey benefits verbally. Send pictures, specifications, and other product literature by mail.
Prove that product works.	Reprints of major press releases, customer testimonials, and ten-day return policy build confidence that the product really works.
Benchmark the performance.	Publish performance charts and benchmark data.
Establish personal rapport.	Increase contact frequency via telephone.
Reduce risk of bad decision.	Ten-day return policy.
Meet decision maker.	Easier to find the manager by phone. He frequently did not attend product demonstrations given to engineers.
Listen for objections.	Trial close on telephone.
Overcome objections.	Immediate access to inside technical support and sales management make rapid response to technical, pricing, configuration, and compatibility issues possible.
Determine if budget available.	Ask if budget available on telephone.
Consult on product application.	Telephone consulting and a crisp, brisk, factual follow-up letter outlining specifics on the prospect's application.
Provide personal service.	Send custom prepared materials such as specific configurations and price quotations immediately in writing via overnight delivery service.

Figure 11–4. Eliminating Costly Field Visits

ganizations generally move slowly and small organizations frequently can move more quickly. Compared to the inherent sales cycle time, none of the delays imposed by delivery times of material or by stretching the accomplishments of a single field visit into several telephone contracts and literature fulfillments are significant.

Sales	Activity	Prospects Remaining	Delivery Vehicle	Elapsed Time (weeks)	Activity Cost	Total Cost
1.	New lead	100	Mail/Telephone	0	$25	$2,500
2.	Initial literature	100	Mail	1	$10	$3,500
3.	Qualify call	100	Telephone	2	$15	$5,000
4.	Second literature	30	Mail	3	$10	$5,300
5.	Consult/Sales call	20	Telephone	4	$25	$5,800
6.	Payback worksheet	10	Telephone	6	$5	$5,850
7.	Configuration call	10	Telephone	6	$10	$5,950
8.	Proposal	5	Overnight service	8	$100	$6,450
9.-12.	Additional follow-up	3	Telephone	12	$100	$6,750
10.	Approval/Purchase	3	Telephone	16	$10	$6,780
14.	Post-sale Follow-up	3	Telephone	20	$30	$6,870

Figure 11–5. Telemarketing-Oriented Sales Process

Risk vs. Efficiency

The telemarketing-oriented sales process clearly conserves marketing resources and is thus more "economical." Critics will argue, and rightly so, that sales can be lost through a less personal process that yields less information and intuitive insight about the prospect. It was established earlier that this will always be true. So why bother saving money selling if you just have a lower probability of success? Well, it just comes down to the numbers. If the new sales process is 95 percent as effective but lets you reach ten times the number of prospects, the benefits are obvious and dramatic. If the sales process effectiveness drops precipitously then one must reevaluate and try again. The trick is to choose telemarketing-oriented marketing-mix substitutions that lower costs, or more importantly, create new sales opportunities to a much greater extent than they negatively affect sale conversion rates.

Yes or No?

By the time you have done your homework you will have developed a feeling for whether telemarketing can help, and perhaps even some specific ideas to try out. If it looks promising, allocate a budget and an adequate amount of your personal time to give it a fair test. Set an overall goal complete with intermediate objectives. Choose modest, easily measured, and simply implemented objectives initially. Try it yourself first. The firsthand experience will be invaluable in selecting and training others to do the work later.

Empirical Iterative Process

Bear in mind that success accrues to the methodical, observant, and patient. Just "making a few calls" or getting the secretary to do it does not constitute a valid trial. Set specific numerical objectives, expend thoughtful energy, measure the results, and compare them to the objectives. If you are off target, reevaluate the objective and the approach in light of the new experience and try again.

There is more to it than just haphazard trail and error. The learning process is quite rapid and after a few days or weeks you may want to reevaluate the whole program based upon the wealth of new information and insight you have gained. Remember that it is an essentially iterative process. The results are easily measurable and the data is purely empirical. Armed with solid numbers, earned by testing theory against reality, one can project the results of a larger-scale rollout. Be careful to test again using the type of person you would hire for the job before extrapolating results based upon only your personal efforts. You may be much better at it (or much worse) than others because of your overall perspective, training, and skills.

Implementation

There are many secondary decisions to make when rolling out the telemarketing effort. These include, but are by no means limited to, selection and training of telemarketers, compensation strategies, telephone equipment selection, territory or account divisions, changes to existing collateral material, and coexistence with current salespeople. Do not fret over all this initially. On a trial basis, you can move right along without having all the details worked out. While success will ultimately depend upon doing these things right, it is important to keep these tactical issues separated from the key strategies being tested.

People Make Sales

It is easy to lose sight in all this analysis that ultimately it is people who make sales. Telemarketing personnel should be chosen with the same care as any other sales person. The requirements of the job are new and somewhat unfamiliar but at a minimum the person should be skilled in the art of selling, have in-depth product knowledge, and have the proper temperament, attitude, and drive to succeed at telemarketing. It may (or may not) be preferable to use people from within the company. Your initial telemarketers should be open-minded, able to readily accept change, unbiased against telemarketing success, and ready for the challenge.

Telemarketing can be successfully used as part of a balanced marketing program to sell high-technology products. The best results are obtained by choosing objectives with the whole marketing effort clearly in view and by deliberately and methodically testing changes in the marketing effort. Understanding of the sales process and the application of the marketing mix are essential. Applicability of telemarketing as a sales tool depends upon a variety of factors and will be unique to each situation.

Potential rewards of telemarketing are fascinating. Through telemarketing, a high-technology product that by its very nature is unprofitable to sell through traditional means may find economical sales channels. Mature product lines in competitive industries may find increased competitiveness through lowered sales costs. Aftermarkets may become more lucrative. Account-opening products can pave the way for more profitable items in the future. Only the bounds of imagination limit telemarketing.

12
Technology Forecasting, New Product Development, and Market Evolution

John H. Vanston
Donna C. L. Prestwood

In a recent article, Joel Goldhar points out the importance in today's "high-tech" environment of coordinating product development and market emergence.[1] Specifically, he says the short commercial lifetimes of many new products today make it imperative that marketing programs begin before the product actually becomes available. Pent-up demand awaiting product availability provides opportunities for very attractive market positioning.

The announcement of the Lotus 1-2-3 software package by the Lotus Development Corporation (often called the worst-kept secret in computer history) was preceded by rumors, leaks, and hints that assured almost instant industry dominance.[2] Likewise, the introduction of the Compaq portable computer was carefully orchestrated to ensure widespread interest,[3] and, more recently, months of customer anticipation for the IBM laptop computer resulted in a rush of orders, reportedly far exceeding original estimates.[4] In some cases, the market need represented "provided" opportunities and, in other cases, "created" opportunities. Regardless of the nature of market development, the key factor of success was the deliberate coincidence of product introduction and market opening.

The concept of defining market windows and developing products to take advantage of these windows is a well-known and often-used market tactic. However, in the present era of explosive technological advance, executing that concept is more difficult and more important than it has been in the past. It is more difficult because the technologies involved are often extremely sophisticated and complex; because multiple, technologically differentiated products exist concurrently in the marketplace; and because typical product life cycles are, indeed, shorter than ever before. It is more important because the investment required to develop, manufacture, and market new products has become so large that missteps often threaten the very survival of a corporation.

A further incentive for better product development/market emergence coordination by American companies is that competitors in other countries have become quite proficient in this art. As the American public, wakened by various consumer groups, became resentful about poor workmanship in U.S.-built automobiles, high-quality Japanese cars arrived on the marketplace. As the price of motion picture tickets skyrocketed and multichanneled cable television increased viewing-time conflicts, the Japanese introduced the videocassette recorder. As audiotape players reached technical limits, the Japanese were ready with new compact disk players. Even as the mechanical parking meter unsuspectingly continues to gobble the coin of the realm, a Swedish company is introducing an electronic meter that automatically adjusts parking fees according to customer load, signals enforcement officers when a car has overstayed its time limits, and responds to tampering with a claxon alarm.

Realizing Potential

The first step in the development of a new product or product line is most often the realization of a potential need for the product. In many cases, organizations wait for this need to manifest itself. Richard Foster, in his new book, *Innovation: The Attacker's Advantage,*[5] illustrates that such waiting often results in the organization's losing its position as a market leader. Foster states that a much sounder policy for companies, particularly leading technology companies, is to continually analyze their products to determine those that are nearing technological limits.

Once such products have been identified, the company should actively search for emerging technologies that might overtake present ones. The histories of American Locomotive, Curtis-Wright, National Cash Register, Addressograph-Multigraph, Mergenthaler, and a host of other companies reflect the danger of "sticking to your knitting" when the market is unraveling.

Obviously, identifying, evaluating, planning, and delivering next-generation products is no easy task. First, one must determine if fundamental technological change is needed and desirable. Second, one must define the technical advances that will make the new product superior to the old and how those advances will affect customers' buying decisions. Third, one must project the time it will take to design, test, and prepare for the manufacture of the new product. Fourth, the market for the new product must be defined and analyzed with particular attention to nonobvious product uses. Product competition—direct, indirect, and structural—must be examined next. Finally, refined product development projections must be made and tested against appropriate contingencies.

For a really new product, accomplishment of these tasks may be complicated by uncertainties in regard to development time, product characteristics, production factors, costs, and so forth. However, these tasks are fundamental to effective product development/market emergence synchronization. It is our belief that the set of approaches, techniques, and application methods commonly known as "technology forecasting" can provide useful assistance in accomplishing each of these tasks, as well as others involved in successful product commercialization.

Although many organizations involved in product development and marketing use one or more of the listed techniques,[6] their application often is not well planned or coordinated. In the following paragraphs, we shall describe how technology forecasting methods can assist in formulating effective new product development/marketing strategies and programs.

Application of Technology Forecasting Techniques

Determination of need for fundamental innovation. Although technology forecasting techniques can and should be used in all phases of the product life cycle and for all types of products, in this article we shall address only premarket entry activities. Further, we shall restrict ourselves to new products that involve one or more fundamental innovations; that is, innovations that provide entirely new functions or perform old functions in a radically different manner. In considering development of such products, company executives must first determine if a "revolutionary" product is necessary or desirable. If technological and market leadership can be maintained by incremental innovation, this approach normally will allow the company to reap maximum advantage from existing capital and intellectual investment.

The decision to deliberately seek obsolescence of a profitable product is an agonizing one, somewhat analogous to a gambler's decision on "when to hold 'em" and "when to fold 'em." The successful executive, as well as the successful gambler, often bases this decision on "gut instinct"—an intuitive reaction backed by a thorough understanding of the underlying probabilities.[7]

As Foster indicates, the time to look for "breakthrough" technologies is when existing technologies are reaching technical limits. He suggests some methods for determining when such limits are being approached (such as talking with technical experts, being alert to missed development deadlines, and noting disagreements between research personnel). But he admits that identification is difficult. This problem is exacerbated by the fact that the limit may be technological, perceptual, or utilitarian. For example, in the early 1940s, the speed of sound was widely considered to be a technological limit for manned aircraft. Obviously, it was not. On the other hand, maximum military aircraft speeds have stayed essentially constant for more than a decade because increased speed no longer offers the utility of increased combat effectiveness.

At this early stage of the analysis process, two technology forecasting techniques can be quite useful in determining if a true limit is being approached: *Exponential Trend Analysis* and *Pearl Curve Analysis*. For an important part of the useful lifetime of many technologies, advances in key technical parameters takes place on a constant percentage, or exponential basis. These advances are indicated by a straight line on semilog graphs.

Using this technique, past progress can be recorded and future progress predicted (see glossary). Most technologies, as they mature, will begin to reach some apparent limit. Often, this limit is approached in accordance with a mathematical formula known as the *Pearl Curve* (see glossary). The shape of the curve can be calculated using the observed rate of exponential advance and the postulated limit.

To determine if a hypothesized limit is a true limit, one can simply examine progress data to see which model, exponential or Pearl, is best reflected. (In the sonic barrier case, for example, the rate of speed improvement showed no sign of "leveling off" as the barrier was reached and surmounted.) If these analyses show that the present technology is indeed nearing a restricting limit, the manager must begin to search for next-generation replacement technologies.

Definition of new product characteristics. A technology forecasting method of particular value in defining the characteristics of the new product is *Morphological Analysis* (see glossary). In using this technique, one begins by listing the major functions that the product must accomplish if it is to achieve its basic purpose. For example, a simple watch must have means for receiving and storing energy, for providing a time passage surrogate, and for transmitting time intelligence to the user. Once necessary functions are defined, different ways of accomplishing each task, including present methods, are listed. Finally, different combinations of proposed methods are examined for utility, practicality, and competitive advantage. Generally when this technique is used, a single limiting function will be identified.

Once transistor technology was available, the key to a successful transistor radio was the development of a small speaker. Japanese companies recognized this and achieved their first commercial electronic breakthrough. The commercial viability of the videocassette recorder was dependent on the development of an inexpensive recorder head and digital watches were made commercially practical by the development of liquid crystal displays.

Accomplishing the Function

Once one or more critical functions have been identified, the manager must examine various ways of accomplishing that function. Usually, a number of methods will have been suggested, ranging from the inspired to the incredible. To select those methods that should be more carefully examined, a variety of *Projective, Trend Analysis,* and *Surveillance Techniques* may be employed (see glossary).

Normally, approaches are selected that would be sufficiently different from present technologies to provide a very significant improvement in attractiveness to customers, but not so different as to provide insurmountable development difficulties. For example, Texas Instruments moved to an early position of dominance in the electronics industry by adopting silicon over germanium as a component foundation. They chose silicon because of its superior electronic properties despite known manufacturing difficulties.[8] They chose not to go to gallium arsenide, which has better properties, but is even harder to manufacture consistently. In selecting technologies to be developed, both *Projective* and *Expert Opinion Forecasts* may prove very helpful (see glossary).

Often, when the method of performing one of the requisite functions is changed, new methods must be developed to perform one or more or the often requisite functions. A few years ago, researchers at the Kodak company sought to decrease the drying time of a certain film type. A *Morphological Analysis* indicated that the radiant heating process being used for drying was the constraining factor in the overall film development process. A forced-air drying process appeared more promising; however, to use this method, a new type of emulsion had to be created. Breakthroughs in both areas made possible today's extremely fast processing services. It should be noted that, as the number of functional areas changed is increased, product development will, in most cases, also increase. However, the probability of major product differentiation will also normally increase as well.

The simultaneous changing of more than a single functional element would increase what Abetti and Stuart refer to as "functional newness" of the product.[9] Although increased functional newness normally increases the risk involved in introducing a new product, Abetti and Stuart demonstrate that the probability of a new product's success is actually increased if the product provides a high degree of innovation uniqueness.

Estimates of the time necessary to bring products into the marketplace. As indicated earlier, there are great advantages in presensitizing potential customers for the introduction of a new product. This presensitizing can be catastrophic, however, if the product is not ready for introduction at the expected time. The Osborne II portable computer was a classic example of this phenomenon. After a year of preparing the market for this revolutionary advance in computer technology, the introduction date slipped three times and the computer itself performed well below expectations.[10] Although many of the device's shortcomings later were overcome, the company's bankruptcy once again illustrated the maxim that "one never gets a second chance to make a first impression."

In order to make preliminary plans for coordinating product marketing and development, a rough estimate of the availability date is necessary early in the analysis process. The *Expert Opinion Methods, Lead-lag Trend Analysis* and *Analogies* with earlier projects can all provide estimates of development time (see glossary). However, products involving fundamental innovations often have very different development patterns than those involving more incremental innovation. Even at this early stage, simple *Mathematical Forecasting Models* may give more reliable estimates than single, overall estimates. One model specifically designed to forecast sophisticated technical developments is the *Partitive Analytical Forecasting*™ method (see glossary). At this point, the logic network for the model can be simple and input time and probability estimates can be somewhat general. Even the rough results thereby obtained should be accurate enough to allow preliminary decisions to be made and planning initiated.

(A consumer product company recently employed an interesting method for applying expert opinion to evaluate a line of products requiring significant technical breakthroughs for introduction. In this case, the project manager conducted *structured interviews* [see glossary] with experts in the appropriate technical areas throughout the world. Estimates of the time necessary to achieve the required technical advances fell into two groups: five to ten years and twenty to thirty years. To resolve this dilemma, the manager contracted with each of the experts in the more optimistic group to prepare a research plan that could result in a breakthrough within the time they estimated. Each of the response plans resulted in a new completion date forecast of at least twenty years!)

Analysis of potential market for new products. Concurrent with the projection of a new product availability date, projections must be made of relevant market factors: when will there be a perceived need for the product? How large will the market be at introduction? How fast and in what manner will the market grow over time? How should pre-entry market strategies be drawn?

Historically, market projections for new products based on fundamentally different technologies have not been particularly good. A market survey before introduction of the Xerox (then Haloid) 914 copier in 1960 projected sales of only five thousand machines per year. Tom Watson, Sr., then president of IBM, reputedly once projected an eventual need for no more than fifty computers in the United States. More recently, Cray predicted a total market for not more than fifty supercomputers by the year 2000; the number of completed and ordered machines already exceeds this number.[11]

The reasons for such erroneous sales forecasts are not difficult to appreciate. The people making these projections simply failed to see the new uses that would develop for the new product, as well as the impact of continuing improvements and advances in the new technology. Market researchers failed to see the impact of an easy-to-use copier on office and administrative procedures; failed to see the dramatic improvement that would occur in computer cost efficiency and capability; failed to appreciate the myriad of scientific, engineering, and human resource areas where supercomputers would allow completely new approaches to sophisticated analysis. Similar myopia[12] is widespread: the inventors of the transistor saw it as a partial replacement of binary vacuum tubes and not as a precursor of integrated circuits. Few predicted that Dr. Land's polarized automobile windshields would lead to an "instant" camera, or that research on superconducting magnets would lead to enhanced clinical diagnostic capabilities through Magnetic Resonance Imaging.

Side-Street Opportunities

Dr. Karl Vesper of the University of Washington calls these unexpected market openings "side-street opportunities," opportunities that only become obvious as one travels the main road of product development. (Sometimes these side streets are actually "on ramps" to "superhighways" of unexpected commercial potential.)

While it never will be possible to identify all the evolutions of a new product or all the uses imaginative customers may devise for it, several technology forecasting methods are available to assist the product manager in projecting market potential. One method is the use of *Nominal Group Conferencing* (see glossary). This method is a structured variation of traditional brainstorming. However, the structure involved materially strengthens the value of the technique for identifying nonobvious product uses. For example, we have used *Nominal Group Conferencing* to assist a metal fabrication company in examining new surface treatment techniques; a major energy company in the design of a coordinated research and development program and a large engineering organization in evaluation of new computerized graphic systems.

The nature of the Nominal group method limits participation to small groups of specialists (five to seven people). A method for involving larger groups of people is the *Delphi Survey* (see glossary). Basically, the Delphi differs from other survey methods in that it involves formal, anonymous interactions between participants. Thus, it is particularly valuable in uncovering unconventional points of view.

Another powerful method for uncovering unexpected market possibilities is the *Impact Wheel* (see glossary). This method causes people to use the left (structural) side of the brain and the right (holistic) side simultaneously. In new product analysis, one might first consider the new capabilities the product will offer. Next, new functions that each new capability might make possible. Finally, specific product ideas that might be undertaken. We have seen this method used by a large pulp and paper industry to determine nontraditional markets for a new paper technology; by a government environmental agency to determine uses for a new computer program; and by a space agency contractor to identify new research projects which might be needed by the government.

Another technology forecasting method for defining the market window for the new product is *Fisher-Pry Substitution Analysis* (see glossary). This method uses insights on the substitution process to project market takeover by a new product. Although, strictly speaking, this method can be employed only after substitution has begun, observation of past takeover rates of similar products often allows the technique to be used to make rough projections, even before the new product is introduced.[13]

Analysis of Competition. Once the potential market situation has been examined, the next step is to analyze the status of possible competition. There will be little satisfaction or gain from developing a robust market for a new product that is first introduced by the competition. There are three types of competition that must be considered: direct competitions—companies that might develop a similar product; indirect competition—companies that might introduce different products that can perform the same function; and structural competition—developments that might eliminate the need for the new product.

Analyzing the overall competitive situation is often the most difficult step in planning new product introductions because the other players in the game will actively seek to hide, withhold, or misrepresent relevant information. Competitive analysis is particularly difficult for products based on fundamentally new technologies because innovations of this type often come outside the traditional industry. Color photography was invented by two musicians; the photocopy process by a patent attorney, and mechanical refrigeration by a medical doctor. Integrated circuits were developed by an oil-well service company; the diesel locomotive by an automobile manufacturer.

Surveillance Techniques

For analyzing direct competition, the surveillance techniques—*scanning, monitoring* and *tracking* can be invaluable (see glossary). In using these techniques, it is normally desirable to sensitize organizational personnel to areas of special interest.

Historically, it has been found that advances in technology often occur almost simultaneously in more than one place. At the time the Wright brothers flew at Kitty Hawk, S.P. Langley was testing a steam-powered plane in Washington, D.C. The jet engine was developed about the same time by Frank Whittle in England and Hans Von Neuman in Germany, without either knowing of the other's work. A hundred years after the basic design of a computer had been produced by Charles Babbage, electromechanical computers were developed simultaneously by Konrad Zuss in Germany, Allan Turning in Great Britain, and Howard Akin in the United States.

Because of this "parallel invention" phenomenon, an organization developing a new product should be apprehensive that other groups are working on a similar product. However, this phenomenon often can be used to the organization's advantage. By examining its own programs, the organization can gain insight on how to best "shadow" those of others. A few year ago, researchers at the Johnson Controls Company were working on an important new process control device. They suspected a competitor was conducting similar research, and therefore, the Johnson people carefully monitored hiring patterns of the competing organization for more than a year. By analyzing these patterns, they were able not only to surmise the progress of the research at the competing firm, but also to discern the technical approach that they were taking. This analysis led to Johnson Controls' product arriving on the marketplace more than six months before the competing product.

As difficult as analysis of direct competition can be, analysis of indirect competition is often harder. In this situation, the morphological analyses conducted earlier can be of significant value. The first step in the previous analysis was to identify all of the reasonably practical ways of accomplishing each major function of the new product. The initial use of the morphological matrix was to decide the approach that your organization would take in new product development. A second examination of the matrix can be used to identify alternate reasonable approaches to product development that other organizations might undertake. This identification can provide you with guidelines for targeting surveillance activities.

Several years ago, a large engine manufacturer planned the introduction of a new family of diesel engines. Company officials were confident that this engine would represent the new state-of-the-art in this field. However, they were concerned some other organizations might be developing a different type of engine that could perform the same functions more efficiently and, thus, erode the market for their engine family. A corporate planning group conducted a morphological analysis to explore this possibility. The results indicated that a new type of engine threat was unlikely, and the engine was successfully introduced.

Perhaps the most difficult but most important type of analysis involves structural changes in the market. Even the most elegant device for accomplishing a given function will be of little value if the need for accomplishing that function is either eliminated or materially reduced. Research on new clothes-ironing devices became essentially worthless with the advent of permanent-press clothes. Development and acceptance of practical discardable clothes would have a similar impact on both the washing machine and detergent industries. There will be little purpose to developing better railroad wheel bearings if magnetic levitation becomes practical and widespread, and development of high speed trains may, in turn, become less important if three-dimension videophones drastically reduce the need for business travel. Often, structural competition can come from a completely unrelated field of technology. Improvements in iron lung technologies, for example, lost most of their consequence with the discovery of the Salk vaccines.

Impact Study

A technology forecasting methodology of particular importance in analyzing structural competition is the *Cross-Impact Matrix* (see glossary). In this technique, the major events that might affect the development and use of the new product are listed, the probability of each event's occurring evaluated, and the effects of each event on other events as well as on the successful commercialization of the new product are assessed. Computer simulations then are used to examine the nature and extent of potential structural interaction.

Among others, the Monsanto Company has used the cross-impact technique for many years to formulate plans for new chemical products. The National Space and Aeronautics Administration used a similar process to evaluate nonhydrocarbon flight fuels; and the Martin-Marietta Corporation has used cross-impact analyses to assess new space program needs.

In structural competition analysis, the use of *Alternate Scenarios* is often also of value (see glossary). In this technique, different sets of assumptions are postulated on how the future might develop. New product plans then are tested against each of these scenarios to determine possible events or trends that might materially affect product success.

Refinement of development projections and translation into project plans.
The final step in planning the coordination of market evolution and product development is a detailed examination of the time required to bring the new product into the market. For this task, product managers may well return to the previously formulated *Partitive Analytical Forecasting*™ network. However, where previously only generally representative logic networks were satisfactory, at this stage one should attempt to make the logic network as representative of the real development process as possible. Where, previously, only rough estimates of time and probabilities were attempted, at this stage these estimates should be as definite and accurate as possible. Experience in the fusion power and the space shuttle programs have shown that careful application of this technique can produce good program projections even for very complex and technically sophisticated projects.

Implementation

In the previous section, we have outlined a structured program for using technology forecasting approaches and techniques for coordinating new product research and development efforts with market development activities during the premarket entry phase of the product life cycle. This technology forecasting program is not intended to stand alone; rather, it should be integrated into other, more traditional analysis and planning approaches.

Each technology forecasting program must be tailored to meet local situational circumstances. The techniques described, for the most part, have easily understood theoretical bases and can be effectively executed by organizational personnel after a relatively small amount of training and experience. The question arises, however, as to how responsibilities can be best allocated for the conduct and application of the appropriate forecasts.

It is apparent that the first two phases of the planning process, the decision to develop a fundamentally new product, and the new product's basic characteristics are of great importance to the success of, even the survival of, the organization. Hence, they should be a matter of major concern to top company management. In a small company, the top executives themselves may well be closely involved in these analyses. In a large company, these responsibilities often will be turned over to a corporate staff group charged with "strategic technology planning." However, key executives should closely monitor the activities of this group.

Many large companies have some type of new venture group responsible for developing new business opportunities. In general, these groups are not appropriate for conducting the types of analyses described, since their focus is normally on new types of business, not on new products designed to replace existing company products.

Once the pivotal decisions of phases one and two have been made, it will be necessary to conduct product development, market, and competition analyses (phase 3–5). For these activities the formation of a task force is often advisable. As William Altier points out, "Task forces are a productive tool to use when resolving decisions that cross functional or organizational boundaries. They are a tool to apply to situations that do not fit neatly into an existing box in the organizational structure."[14] The make-up of the task force will depend on the specific product and organization involved. However, representatives of the appropriate product development and marketing functions obviously will be included.

As new product planning goes forward, a more permanent organizational structure will be necessary. The group given final responsibility for managing the new product should also be charged with conducting the technology forecasting efforts involved in final product planning. In most cases, this new organization should be independent from the group responsibility for the product being replaced. Arnold Cooper, among others, has demonstrated that there are many difficulties involved in having the same organization manage both the replacing and the replaced products and technologies.[15]

Although maintenance of continuing profits is a very important part of top management's responsibility to the company, the long-term success and survivability of the organization is even more important. To achieve both these goals, it is essential that top management companies examine the long-term viability of current major product lines, while concurrently searching for totally new product opportunities. A well-conceived and carefully executed technology forecasting program, such as the one described above, can contribute materially to the successful accomplishment of these endeavors.

Notes

1. Joel D. Goldhar, "Informal Notes on Organizing and Managing the R&D/Marketing Interface for Innovation," presented at the Teletype Corporation Seminar on Management of R&D, January 30, 1985.

2. Regis McKenna, *The Regis Touch* (Reading, Mass.: Addison-Wesley, 1985), 67.

3. Rod Canion, "America's Fastest Growing Company: Compaq's Market Creation Strategy," *High Technology Marketing Review*, Spring 1987, 1–8.

4. "Guess What Computer Makers Want for Christmas," *Business Week*, October 14, 1985, 44.

5. Richard N. Foster, *Innovation: The Attacker's Advantage* (New York: Summit Books, 1986).

6. William L. Shanklin and John K. Ryans, Jr., *Marketing High Technology* (Lexington, Mass.: Lexington Books, 1984).

7. Weston H. Agor, *Intuitive Management* (Englewood Cliffs, New Jersey: Prentice-Hall, 1984).

8. T.R. Reid, *The Chip: The Microelectronics Revolution and the Man Who Made It* (New York: Simon and Schuster, 1985).

9. Pier A. Abetti and Robert W. Stuart, "New Product Development Through Technological Innovation and Entrepreneurship," presented at the Conference on "Entrepreneurial Management: Innovation and New Market Development," at the IC^2 Institute, the University of Texas at Austin, Texas, April 18-19, 1986.

10. *Wall Street Journal,* February 6, 1986, 30.

11. *Wall Street Journal,* February 19, 1986.

12. Theodore Levitt, "Marketing Myopia." *Harvard Business Review,* September/October 1975.

13. Ralph C. Lenz, "Rates of Adoption/Substitution in Technological Change," monograph, published by Technology Futures, Inc., Austin, Texas, 1986.

14. William J. Altier, "Task Forces—an Effective Management Tool," *Sloan Management Review,* Spring 1986, 69–76.

15. Arnold C. Cooper and Dan Schendel, "Strategic Responses to Technology Threats," in *Readings in the Management of Innovation,* Michael L. Tushman and William L. Moore (eds.) (Boston, Mass.: Pitman, 1982), 325–334.

Glossary of Technology Forecasting Terms

Technology forecasting describes a group of techniques that predict in quantifiable terms the direction, character, rate, implications, and impacts of technical advance. TF techniques are based on the logical treatment of credible data and should produce results that are informative and independent of the analyst performing the forecast. Although TFs can serve as effective devices for communication of ideas or for gaining insight on technical progress, they are primarily conducted to provide information to assist planners and managers in making better decisions.

The technology forecasting techniques may be classified into five categories: Surveillance, Projective, Normative (Goal-Oriented), Expert Opinion, and Integrative. Each category and its associated techniques are described below.

Surveillance Techniques

Scanning, monitoring, and tracking—are all essentially passive or observational techniques. In many ways, they are similar, differing primarily in the degree of focus and intensity of the information search effort. Although application of the three techniques may overlap and the same people may be involved in all three simultaneously, it is desirable to keep in mind the different nature of each technique.

Scanning is the term applied to broadly oriented surveillance activities that seek to identify, at an early time, developments in technical, economic, social, political, and ecological environments that may materially affect the organization. Scanning programs are particularly useful in identifying emerging products and processes that provide commercial advantage to the organization or its competitors, in pointing out new uses for present or developing technologies, and in suggesting sociopolitical factors that impede or enhance acceptance of new technologies.

Monitoring implies a more focused and disciplined type of surveillance effort than scanning. An effective monitoring plan should include five basic elements: (1) a system for selecting areas to be monitored; (2) a system for assigning specific monitoring responsibilities to individuals or groups; (3) an assessment of progress levels at which developments will become significant to the organization; (4) a clear definition of the action the responsible agent is to take when progress reaches the designated level of importance; and (5) a formal procedure for periodically reviewing the monitoring program and related databases.

Tracking is the most carefully focused and intensive of the surveillance techniques. It involves a very concentrated effort to follow developments of major significance to the organization—competitive response to a new product or process nearing introduction; possible near-term introduction of new products and processes that will threaten present markets, and major technical breakthroughs in which the organization needs to establish a position. Often, special *ad hoc* groups are organized to conduct or supervise specific tracking efforts.

Projected Techniques

The projected techniques are based on the theory that, for some period of time, the future will be very much like the past. It assumes that events, trends, and development patterns in the past were shaped by various fundamental driving forces and that, as long as these forces do not change significantly, past patterns of change will continue in the future.

Technical Trend Extrapolation. A large body of empirical data indicates that when the values of key parameters of technical progress are plotted against time, a regular development pattern can be discerned. In a large number of technical areas, it has been found that if progress is plotted versus time, the trace is linear on semilog graph for a significant portion of the development period. These projections can be very useful in setting organizational R&D goals, in estimating the progress that may be taking place outside the organization, and in identifying the need for new technical approaches.

Pearl Curve. As a technology matures, it will almost invariably begin to approach limits to its development. These limits may be either real or perceived. As a limit is neared, exponential improvement is slowed; thus, it is often useful to know how the limit will be approached. In many cases, it has been found that technical progress in this part of the development period can be approximated by the Pearl formula:

$$y = \frac{L}{1 = a \exp(-bt)}$$

Where, y = value of the parameter at time t, L = parameter limit and a and b are constants determined from known data.

Precursor Developments. It has been observed over a period of years that technical development in one area follows development in other areas in a predictable manner. When such lead-lag relations exist, it is often possible to forecast developments in the lagging technical area by observing the state of development in the leading one. Where progress in the leading technology can also be forecast, this allows reasonable extension of the forecasts' time horizon for the lagging technology.

Substitution Analysis. When a given technology begins to mature (that is, as improvements become increasingly difficult and expensive), a new technology often will emerge that can accomplish the required function in a more effective and economic manner.

Development of the new technology will allow it to take over progressively larger segments of the market. Analysis of this substitution phenomena indicates a general pattern to this process, and a number of TF practitioners have described it in mathematical terms. One of the more successful of these attempts was a substitution formula developed by John Fisher and Robert Pry:[1]

$$\frac{f}{1 - f} = a \exp(bt)$$

Where, f = degree of substitution at time t and a and b are empirically derived constants. This formula has been found to roughly describe substitution histories in a number of technical developments in a wide variety of fields. When used for forecasting purposes, one can use early substitution data to determine constants and then project the fraction of market takeover at any time in the future.

Normative (Goal-Oriented) Techniques

While projective forecasting focuses on understanding the past, normative (goal-oriented) forecasting is based on the assumption that future technology development will be driven by future needs. It assumes that, as future needs are perceived, society will make available sufficient funds, facilities, and people to develop means of satisfying those needs. Thus, if one can identify the perceived needs of society—or a significant portion of that society—in the future, one will be able to reasonably forecast the technologies that will be developed to satisfy those needs.

The practical use of normative forecasting involves three tasks: (1) identifying societal needs; (2) identifying technologies that satisfy those needs; and (3) selecting those new technologies that best coincide with the organization's goals, capabilities, and competitive status.

Impact Wheel is a technique for using an informed panel to identify higher order, often nonobvious, effects, and implications of selected decisions or developments.[2] Use of the technique starts with specification of an event, trend, technical advance, or societal development one wishes to analyze. The panel is then asked to identify the direct consequences of the occurrence of this central item. Once five to seven direct consequences are identified, the panel is asked to suggest possible implications that might arise from each of these first-order consequences. This process is continued for third and higher-order consequences to the extent that it is useful.

Experience has shown that this technique can be a very potent means of identifying unexpected or easily overlooked opportunities, problems, and interrelationships.

Morphological Analysis. Most new technologies grow out of a desire to perform the technical function more efficiently or economically. However, in most modern equipment and systems, many subordinate functions are involved in the accomplishment of the overall function. In morphological analysis, the forecaster determines what the more subordinate functions are; identifies the methods that are being used in present systems to accomplish those functions; identifies alternate means of accomplishing each function; and examines different ways of combining the subordinate technologies to suggest innovative approaches to accomplishing the basic functions of the overall system. Experience has shown that this technique is useful in spurring imaginative ideas about new ways of combining the subordinate technologies to suggest innovative approaches to accomplishing the basic functions of the overall system. Experience has shown that this technique also is useful in spurring imaginative ideas about new ways of meeting both existing and postulated future needs. It can be a valuable device for defensive forecasting (that is, for providing insights into competing products or processes that might be under development).

Relevance Trees. For the results of a normative forecast to be applied, the organization must select those technologies that coincide with its objectives and capabilities. One device for assistance in this selection process is the relevance tree technique. The basic principle of this methodology is the division of the relevant elements of a decision into increasingly smaller components; the establishment of formal criteria for specifying the relative importance of each component; the assignment of numerical values for each component and criteria; the quantitative rates of alternate technical solutions against each criteria and component; and, the combination of individual ratings to give new insights on the overall decision.

Both the advantages and disadvantages of the method lie in its simplicity. Often, the most important outcome of relevance three application is the clarification of relationships between factors and the identification of areas requiring more careful study.

Expert Opinion

In an ongoing, technically oriented company, there resides a reservoir of technical talent, experience, and training—scientists, engineers, salespeople, technicians, and so on. A well-formulated technology forecasting program will take advantage of the knowledge and wisdom of this collection of experts.

Delphi Surveys. The Delphi procedure involves four specific rounds. In the first round, the experts are asked to estimate when they expect each of a number of events to occur.[3] When answers are received they are tabulated and, in round 2, sent to respondents who are requested to reconsider their original projections. Respondents whose Round 2 answers fall in the upper or lower Round 1 quartiles are requested to provide reasons for their estimates. In the third round respondents are sent the retabulated totals, together with the nonattributed comments gathered in Round 2. Respondents are requested to once again consider their previous projections and are invited to add comments, if desired. For the final round, the person conducting the survey tabulates and distributes all projections and comments. The results of a Delphi survey may be useful in a number of ways. The mean value of estimates gives a projection of when a given technical development may be expected to occur, while the spread of these values can give an indication of the degree of agreement between experts. The survey also may be used to compare the estimates of one group of participants with other groups.

Structured Interviews. The structured interview technique is conceptually very similar to the Delphi survey. In this technique, the person conducting the survey personally collects data and acts as an intermediary of ideas. As opposed to procedures for free-form interviews, subject introduction techniques, sequence and nature of questions, discussion procedures, and all administrative details are standardized before initiation of the project. The interviewer goes from one participant to the next adding each interview's results to the data base for the next interview. When all interviews have been conducted, feedback is completed by telephone or written communication with earlier participants.

Nominal Group Conferencing is designed to improve the use of expert opinion.[4] It is most effective when a small panel of experts, five to seven, is available for about half a day.

Experience has shown that nominal group conferencing can be a very useful tool for eliciting potential and imaginative ideas from a group of experts. The method almost assures that all members actively participate in the conference and minimizes many of the social dynamic problems associated with committee meetings. Often, it is useful to conduct a series of nominal group sessions on the same subject matter and analyze differences and similarities in results.

Integrative Techniques

Technical advance does not take place in a vacuum. New technologies are often triggered or accelerated by advances in an entirely different technology. Technical advance is also often enhanced or deterred by nontechnical factors. Thus, forecasts about future development of a technology must not only take other technical and nontechnical developments into account, but must also specify and evaluate these relationships. A number of technology forecasting techniques have been devised for analyzing the effects of various development factors on one another and for presenting decision makers with an integrated picture of how technical process may take place.

Cross-Impact Analysis. Although all technology forecasting methods take implicit cognizance of exogenous influences, it is necessary, on occasion, to address these influences explicitly. One method for taking formal account of such interactive effects is the cross-impact analysis technique. Typically, such analyses begin with an identification of those factors that will most significantly affect the technical development being considered or which will be most affected by the development. These effects might include a change in the probability of the event occurring, of the time of its occurrence, or of the significance of the development. Once the factors to be considered are determined, they are arranged in a matrix format with the basic technical development and the other factors arranged in the left-hand column of the matrix and the same arrangement used to form an equal number of columns at the top. The people conducting the analysis complete the matrix by placing in each box a notation of the effect of the occurrence of each element in the left-hand column on each of the other elements of the matrix.

To simultaneously consider a large number of interactions, mathematical formulas relating events are often developed. With appropriate input data, these formulas result in projections that take into account all included factors. Because solution of the relevant equations may be difficult, computers are often used.

Cross-impact analysis can be a very effective method for examining the probability that a technical development will be affected by exogenous factors. Thus, it can provide a basis for a targeted surveillance program.

Scenarios. One method of examining and presenting the interactions between projections of a number of technical and nontechnical factors is to combine them into a integrated description of the future.[5] Such descriptions are often referred to as "scenarios" and are quite useful in technology planning. Because a scenario presents a multifaceted portrait of the future, it allows more realistic consideration of "real world" situations and adds both breadth and depth to decisions bout future operations. Moreover, because of its "story" orientation, it often allows the organization to consider alternative futures in a serious, but nonthreatening manner.

The use of alternate scenarios in planning can involve considerable expenditure of time and effort. However, experience has shown that the technique illustrates for planners the importance of flexible planning; serves as an excellent interorganizational communication tool; provides a vehicle for integrating relevant technical and nontechnical factors in the planning process; provides a basis for an effective monitoring plan; and identifies important decisions that will have to be made in the future.

Mathematical Models. Application of most of the technology forecasting methods described above can be abetted by the use of simple mathematical models. It helps to use the power of modern computers to integrate a number of technical and nontechnical projections and to take into account the interactions between them. Use of appropriate mathematical models and computer simulations may allow the consideration of many more factors than might otherwise be possible and may also permit the planner to test the implications of different organizational programs and strategies.

Partitive Analytical Forecasting.™ In this model, all tasks necessary for product introduction are identified; different methods for accomplishing these tasks are determined; and the relationship between tasks and means of accomplishment are integrated into a development logic network similar to a Program Evaluation and Review Technique (PERT) network.[6] Estimates of the time necessary to complete each task and probability of complete or partial task completion are determined using expert opinion, past experience, or trend analysis. Finally, computer simulation is used to determine the probability of product development success as a function of time.

Notes

1. John C. Fisher and Robert R. Pry, "A Simple Substitution Model of Technological Change," Report No. 70-C-215 (Schenectady, N.Y.: General Electric Company, June 1970). Also appears in M.J. Cetron (ed.), *Industrial Applications of Technology Forecasting* (New York: John Wiley & Sons, 1971). For additional in-

formation on Fisher-Pry substitution model, contact Ralph C. Lenz, Technology Futures, Inc., 6034 West Courtyard Drive, Suite 380, Austin, Texas 78730.

2. For additional information on impact wheels, contact Joel Barker, President, Infinity Limited, Inc., 1301 Cherokee Avenue, West St. Paul, Minn., 55118.

3. For further discussion of the Delphi method, see Harold A. Linstone and Murray Turnoff (eds.) *The Delphi Method, Techniques and Applications* (Reading, Mass.: Addison-Wesley, 1975).

4. For further discussion of the nominal group conference technique, see A.L. Delbec, A.H. Van de Ven, and D.H. Gustafson, *Group Techniques for Program Planning* (Glenview, Ill.: Scott-Foresman, 1975).

5. For additional information concerning the structure and use of scenarios, see John H. Vanston, et al., "Alternate Scenario Planning," *Technology Forecasting and Social Change* 10 (1977), 159–180.

6. For additional information, see John H. Vanston, et al., "PAF—A New Probabilistic, Computer-Based Technique for Technology Forecasting," *Technology Forecasting and Social Change* 10 (1977), 239–258.

Index

ABA (American Bar Association), 61
Abbott-Abbott, Edwin, 3
Abernathy, William J., 42
Abetti, Pier A., 163
Adams; A., 21–23
Addressograph-Multigraph, 160
Akin, Howard, 167
ALCO, 25
Allen, Gracie, 7
Alternate scenarios, 168, 179
Altier, William, 169
AMA (American Medical Association), 61
Amdahl, 35
American Airlines, 87
American Bar Association (ABA), 61
American Home Builders Association, 69
American Locomotive, 160
American Medical Association (AMA), 61
AMP Corporation, 69
Apple Computer, 34; Apple II, 4–5, 8; Apple III, 6; Compaq versus, 53; image of, 76; industry standards and, 130; innovation at, 87; Macintosh, 5
Ashton-Tate: Bottom-Line Strategist, 113; case study of, 109–110, 112–118; ChartMaster, 109; company profile of, 109–110; corporate objective of, 115; dBASE, 110, 111, 113; dBASE II, 114; dBASE III Plus, 109; Decision Resources acquired by, 109; Diagram, 109; family culture at, 112–113; Financial Planning Language, 113; founding family unhappy at, 116–117; Framework II, 109; Friday!, 113; future of, 117–118; garage-shop era at, 110, 112; growing-corporation era at, 110, 117; image of, 115–116; lord-and-master era at, 113–114; Map, 109; Multimate International acquired by, 109; organized-business era at, 110, 114–115; ruling-prince era at, 110, 113; SignMaster, 109; stages of growth at, 110
AT, IBM-PC, 52
AT&T, 69, 70; Compaq versus, 51, 52; divestiture of, 62, 71; personal computers, 51, 52; telephone switching equipment, 59; videotex, 74, 75
Atari, 10; decentralization of, 87–88

Babbage, Charles, 167
Bell Laboratories, 59
Bell local telephone companies, 62, 71
Booz-Allen and Hamilton, 19–20, 57, 58
Boston University, 121
Bottom-Line Strategist (software program), 113
Brazil: high-tech manufacturing in, xv; technological innovation in, 57
Britain: as exporter, xiv-xv; Industrial Revolution in, xiv-xv; marketing in, 26; marketing study in, 21–22
Brooks Brothers, xviii
Burns, George, 7

CAD/CAM (computer-aided design/computer-aided manufacturing), 92
Campbell Soup Company, 87
Canada: marketing study in, 20, 21; technological innovation in, 29
CAT scanners, 25, 27
Cavanaugh, Richard, 12
CCIT (Consultative Committee on International Telecommunications and Telegraphs), 70
Centrex switching systems, 63
ChartMaster, 109
China, People's Republic of: high-tech manufacturing in, xv; technological innovation in, 57
Clifford, Donald, 12
Compaq Computer Corporation: Apple versus, 53; AT&T versus, 51, 52; case study of, 45–53; competitiveness of, 51; DeskPro, 52; IBM versus, 51–52; key factors in success of, 50–51; major beliefs of, 46–47; needs, fulfilling customers' 47–49; portable computer, 47–48, 50, 51, 159; product strategy of, 52; top priorities of, 52–53; turning point in growth of, 49–50
Computer-aided design/computer-aided manufacturing (CAD/CAM), 92
Computer Integrated Manufacturing Systems, 92, 121
ComputerLand, 39, 49
Consultative Committee on International

Telecommunications and Telegraphs (CCITT), 70
Cooper, Arnold, 170
Cooper, R.G., 20, 22
Crawford, C.M., 20
Cray Research, 164
Cross-impact analysis, 168; detailed definition of, 178–179
Curtiss-Wright, 160

Data General, 74
Datamation, xx
Dataquest, 35
dBASE, 110, 111, 113; dBASE II, 114; dBASE III Plus, 109
DEC. *See* Digital Equipment Corporation
Decision Resources, 109
Deforest, Lee, 81
Delphi surveys, 165; detailed definition of, 177
DeskPro, Compaq, 52
Diagram (software program), 109
Digital Equipment Corporation (DEC), xx, 37, 74; expert systems, 127; VAX, 4–5, 127

8048 microprocessor, xix
8080 microprocessor, xix, 5
8086 microprocessor, xix, 5, 38
80386 microprocessor, 38
Eastman Kodak, 163
EHV, Project, 29
Einstein, Albert, 81
Electronic News, xx
Electronic Trend Publications, 38
Electronics, xx
Emerson, Ralph Waldo, 84
EMI CAT scanner, 25
Emycin, 127
Environmental Protection Agency (EPA), 62
Esber, Ed, 114–116
Executive Line, 38
Expert opinion, 163, 164; glossary of, 177–178
Exponential trend analysis, 162

Fairchild Semiconductor, 10; 3870 microprocessor, xix
Federal Trade Commission (FTC), 62
Financial Planning Language (software program), 113
Fisher, John, 175
Fisher-Pry Substitution Analysis, 166, 175
Flatland (Abbott-Abbott), 3
Ford, Henry, 84
Ford Motor Company, xvi, 13
Fortune: innovation examples cited by, 87; *Fortune* 500, 12, 45, 86–87

Foster, Richard, 160, 161
Fourth Dimension, The: Toward a Geometry of Higher Reality (Rucker), 3, 4
Framework II, 109
France: as importer, xv; marketing in, 26
Friday! (software program), 113
FTC (Federal Trade Commission), 62

Gates, Bill, 115
Geneen, Harold, 86–87
General Electric (GE), 17, 69; CAT scanners, 25, 27; data processing systems, 26; image of, 75, 76; innovation at, 87; MRI diagnostic systems, 27; Project EHV, 29; service business, 25–26, 29
General Motors (GM): diesel electric locomotives, 25; "factory of the future," 121; robotics, 130
Germany, as importer, xv. *See also* West Germany
GLisp, 127
GM. *See* General Motors
Goal-oriented (normative) techniques, glossary of, 175–177
Goldhar, Joel, 159
Goldstar, 76
Grid Computer, 38
Griliches, Z., 24
GTE, 61

Haloid, 164
Harris, Jim, 47
Hayes, Robert H., 42
Hewlett-Packard, xvi; R&D at, 33–34, 88

Iacocca, Lee, 86
IBM, xviii, 10, 11, 14, 16, 39, 45, 64–65, 72, 75, 164; 3090 class computers, 36; Compaq versus, 51–52; decentralization of, 87; image of, 76; industry standards and, 130; laptop computer, 159; PC, 3–6, 8, 9, 31, 47–48, 68, 69; PC AT, 52; PC Jr. ("Peanut"), 6, 31, 87; photocopiers, 59; portable computer, 51; ROLM Corporation acquired by, 69; service support, 74; System 360, 35
Impact wheel, 165–166; detailed definition of, 176
India: high-tech manufacturing in, xv; technological innovation in, 57
Innovation: The Attacker's Advantage (Foster), 160
Inside the Black Box: Technology and the Economy (Rosenberg), 11
Integrated Services Digital Network (ISDN), 60, 66–67, 70, 71
Integrative techniques, glossary of, 178–179
Intel, xviii-xix, 40; 8048 microprocessor,

xix; 8080 microprocessor, xix, 5; 8086 microprocessor, xix, 5, 38; 80386 microprocessor, 38; innovation at, 87
ISDN (Integrated Services Digital Network), 60, 66–67, 70, 71
Islam, innovation and, 75
ITT, 86; telephone switching equipment, 59, 61

Japan: as a closed market, 13; as exporter, xv; high-tech manufacturing in, xvi, xx, 70; marketing success of, 11, 39, 65, 70, 79, 160; Ministry of International Trade and Industry (MITI), 70; technological innovation in, 61, 70, 79, 160, 162
Jobs, Steve, 6, 76
Johnson Controls Company, 167

Kapor, Mitch, 115
KEE, 127
Korea. *See* South Korea

Land, Edwin, 165
Langley, S.P., 167
Lashlee, Hal, 110
Lead-lag trend analysis, 164, 175
Leahy, William, 81
Literary Digest, The, 81
Arthur D. Little, 59
Lotus Development Corporation, xviii, 115; 1–2–3, 4–5, 114, 159

Macintosh, Apple, 5
Magnetic Resonance Imaging (MRI) Systems, 27, 165
Magnuson Computer Systems, 36
Managing (Geneen), 86–87
"Managing Our Way to Economic Decline" (Hayes and Abernathy), 42
Mansfield, E., 20
Map (software program), 109
Marketing, definitions of, xvii–xviii
Marketing High Technology (Shanklin and Ryans), 80
Martin Marietta Corporation, 168
Mathematical models, 164; detailed definition of, 179
Memorex, 36, 42
Merck, 87
Mergenthaler, 160
Microsoft, 115
Monitoring, 166; detailed definition of, 174
Monsanto Company, 168
Morgan, James, 87–88
Morita, Akio, 79
Morphological analysis, 162, 163; detailed definition of, 176
Mostek, 48

Motorola, 11; 6800 microprocessor, xix; 68000 microprocessor, xix; 68030 microprocessor, 38
MRI (Magnetic Resonance Imaging) Systems, 27, 165
Multimate International, 109
Murto, Bill, 47–49

NAS (National Advanced Systems), 35
NASA (National Aeronautics and Space Administration), 168
National 16000 microprocessor, 38
National Advanced Systems (NAS), 35
National Aeronautics and Space Administration (NASA), 168
National Cash Register, 160
Netherlands, as importer, xv
Next American Frontier, The (Reich), xiv–xv
Nike, 37
Nominal group conferencing, 165; detailed definition of, 177–178
Normative (goal-oriented) techniques, glossary of, 175–177
Northern Telecom, 59

1–2–3, Lotus, 4–5, 114, 159
OPS–5, 127
Osborne Computer Company, 14; bankruptcy of, 163–164; Executive Line, 38; Osborne I, 38; Osborne II, 163–164

Partitive Analytical Forecasting.MDSU/TM, 164, 168; detailed definition of, 179
PASCAL, 127
PBX switching systems, 59–61, 63, 71
PC, IBM, 3–6, 8, 9, 31, 47–48, 68, 69
PC AT, IBM-, 52
PC Jr. ("Peanut"), IBM, 6, 31, 87
Pearl curve, 162; detailed definition of, 174–175
Pepsico, 34
PERT (Program Evaluation and Review Technique), 179
Peters, Thomas J., 39, 68
Philip Morris, 87
Polaroid, 8
Porter, Michael E., 39
Prime Computer, 74
Product, definitions of, xviii–xx, 15
Program Evaluation and Review Technique (PERT), 179
Project EHV, 29
Projected techniques, 163; glossary of, 174–175
Pry, Robert, 176

Quarterly Review, The, 80

Raster Technologies, 30–31
Ratliff, Wayne, 110
RCA, 10, 69; televisions, 70, 81; videodisc players, 65
Reich, Robert, xiv–xv
Relevance trees, detailed definition of, 176–177
Rensselaer Polytechnic Institute (RPI) Incubator, 30
Riggs, Henry, 34
ROLM Corporation, 69
Rosen, Ben, 48
Rosenberg, Nathan, 11
RPI (Rensselaer Polytechnic Institute) Incubator, 30
Rucker, Rudy, 3, 4

6502 microprocessor, 5
6800 microprocessor, xix
16000 microprocessor, 38
68000 microprocessor, xix
68030 microprocessor, 38
Salk vaccines, 168
Santa Clara University, 40, 42
Savin, 75
Scanning, 166; detailed definition of, 173–174
Scenarios: alternate, 168, 179; detailed definition of, 179
Sculley, John, 34
Sears Business Systems, 49
Sevin, L.J., 48
Siemens, 59, 61
SignMaster, 109
Sony, 79; VCRs, 70
South Korea: as exporter, xv; high-tech manufacturing in, xv; negative image of products made by, 76
Spain, marketing in, 26
Star, Xerox, 5
Storage Technology, 36
Structured interviews, 164; detailed definition of, 177
Stuart, Robert W., 163
Substitution analysis, 175; Fisher-Pry Substitution Analysis, 166, 175
Surveillance techniques, 163; glossary of, 173–174
Symbolics, 127
System 360, 35

3M, 13, 69; innovation at, 87
360, System, 35
3090 class computers, 36
3870 microprocessor, xix
Taiwan: as exporter, xv; high-tech manufacturing in, xv

Tate, George, 110, 114
Technical trend extrapolation, detailed definition of, 174
Technological innovation, definition of, 22
Technology forecasting terms, glossary of, 173–179
Texas Instruments, 39; decision to use silicon, 163
Tracking, 166; detailed definition of, 174
Trend analysis, 163
Trilogy Systems, 36
Truman, Harry, 81
Turning, Allan, 167
Two-Pi Corp., 36

Unisys, 42
United States: business schools in, 42; computer dealers in, 49–50; computer literacy in, 75; economy of, 109; as exporter, xv; government regulation in, 62, 71; high-tech manufacturing in, xvi, 3, 91; as importer, xv; marketing in, 26, 33–40, 42–44, 75, 91, 159–160; marketing studies in, 40, 42, 57, 58; Silicon Valley lessons, 33–40, 42–44; technological innovation in, 29, 61, 66–72, 75, 91, 159–160. *See also individual companies and U.S. government agencies*
U.S. Department of Justice, 62
University of Washington, 165

VAX, 4–5, 127
Vesper, Karl, 165
Victor, 14
VisiCalc, 5
Visicorp, 114
Von Neuman, Hans, 167
Vulcan (software program), 110

Waterman, Robert H., 39, 68
Watson, Tom, Sr., 164
Wells, H.G., 81
West Germany, high-tech manufacturing in, xvi
Whirlpool, 69, 70
Whittle, Frank, 167
Winchester Systems, 147, 150, 152, 153
Winning Performance, The (Clifford and Cavanaugh), 12
Wozniak, Steve, 6
Wright brothers, 167

Xerox, 10, 37, 75; 914 copier, 164; computers, 59, 127; expert systems, 127; Star, 5

Zuss, Konrad, 167

About the Contributors

Pier A. Abetti is professor of Management of Technology in the School of Management at Rensselaer Polytechnic Institute. From 1948 to 1981, Dr. Abetti worked for General Electric in the design of power transformers and in computer systems management. He also has been deputy general manager of UNIVAC in Europe. His present interests are focused on strategic business planning for high-tech corporations, and management of technological innovation and entrepreneurship.

Albert V. Bruno (Ph.D., Krannert School, Purdue) is the Glenn Klimek Professor of Marketing at the Leavey School of Business, Santa Clara University. He is currently director of the Technology Business Research Center at Santa Clara.

Rod Canion is co-founder, president and chief executive officer of Compaq Computer Corporation, one of the leading portable computer manufacturers. He has directed Compaq since its founding in February 1982 and has been responsible for its record-setting growth in all operational areas. His background includes more than sixteen years of engineering, management, and entrepreneurial experience.

William H. Davidow is a general partner with Mohr Davidow Ventures in Menlo Park, California. Before forming this venture capital firm, he was senior vice president of sales and marketing for Intel Corporation and shepherded the renowned Intel 8080 and 8086 to success. Before joining Intel he was a marketing manager for Hewlett-Packard's computer group. He graduated summa cum laude from Dartmouth College and holds a Ph.D. in electrical engineering from Stanford University.

Edward M. Esber, Jr. is chairman and chief executive officer of Ashton-Tate. He has held key managerial positions with several major companies, including IBM and Texas Instruments. He holds an M.S.E.E. from Syracuse University, a B.S. in computer engineering from Case Institute of Technology, and an M.B.A. from Harvard.

Peter LaPlaca is an associate professor of marketing at the University of Connecticut. He received his Ph.D. from Rensselear Polytechnic Institute. He has been studying high-technology marketing for five years and is researching marketing approaches for high-technology products.

Joel Leider is president, vice-president of marketing and co-founder of Winchester Systems, Inc., formed in 1981. He holds a B.S. in electrical engineering from New York University, an M.S. in computer science from Rensselear Polytechnic Institute and an M.B.A. in marketing and finance from Boston University. Before 1981, he was engineering manager at Teradyne, Inc. where he worked for eight years.

Dennis Lewis is co-founder and president of Hi-Tech Public Relations, Inc., a San Francisco-based firm specializing in business, industrial, and consumer public relations for high-technology firms. He has written numerous articles for technical, business, and professional publications and is co-publisher of *Computer Publicity News,* a newsletter for marketing professionals in the computer industry.

Regis McKenna is chairman of Regis McKenna, Inc., an international marketing consulting firm based in Palo Alto, California. His clients include Apple Computer, Lotus Corporation, and many other leading technology firms. He has been a member of the Berkeley Roundtable on International Economy and president of the American Competitiveness Council. He is author of *The Regis Touch: Million-Dollar Advice from America's Top Marketing Consultant* (Addison-Wesley 1985) and *Who's Afraid of Big Blue? How Companies Are Challenging IBM and Winning* (Addison-Wesley 1989).

Donna C.L. Prestwood is a senior associate and director of marketing for Technology Futures, Inc. She has been active in the technology forecasting field for several years. She has managed forecasting projects and participated in courses in technology forecasting and planning for major companies in the United States, Canada, Mexico, and Europe.

Girish Punj is an associate professor of marketing at the University of Connecticut and is interested in the export of technology products. He received his Ph.D. from Carnegie Mellon University.

S. Ram is assistant professor of marketing at the University of Arizona, Tucson. He obtained his Ph.D. in marketing at the University of Illinois at

Urbana-Champaign. Ram's research and teaching interests are primarily in the area of product and service innovations. His papers have appeared in several marketing journals and he is the co-author, along with Jagdish Sheth, of a book titled *Bringing Innovation to Market: How to Break Corporate and Customer Barriers,* (John Wiley and Sons 1987).

John K. Ryans Jr., is a professor of marketing and international business at the Graduate School of Management, Kent State University, Kent, Ohio. He is co-author of *Marketing High Technology* (Lexington Books, D.C. Heath & Company, Lexington, Mass., 1984), and *Essentials of Marketing High Technology* (Lexington Books, D.C. Heath & Company, 1987), as well as numerous articles on the topic.

William Shanklin is professor of marketing at the Graduate School of Management, Kent State University, Kent, Ohio. He is co-author of *Marketing High Technology* (Lexington Books, D.C. Heath & Company, Lexington, Mass., 1984), and *Essentials of Marketing High Technology* (Lexington Books, D.C. Heath & Company, 1987), as well as numerous articles on the topic.

Jagdish Sheth is Robert E. Brooker Distinguished Professor of Marketing at the University of Southern California and has served as a consultant to numerous industrial and consumer products companies including AT&T, General Motors, Whirlpool, and Rockwell International. Dr. Sheth is the author of many books, including *Winning Back Your Market: The Inside Stories of the Companies that Did It.*

Rajendra K. Srivastava is professor of marketing and the Charles A. Le-Maistre Centennial Fellow, IC² Institute at the University of Texas at Austin. He received his Ph.D. from the University of Pittsburgh. His research focuses on competitive market strategies and the evolution of technological markets.

Michael Stone is a marketing consultant in California. He was director of New Products, Research, and Planning for Ashton-Tate. Before that he was with McKinsey and Company and Future Computing as a consultant working with packaged goods and high-technology companies. He holds a B.B.A. and M.B.A. from the University of Texas at Austin.

Robert W. Stuart is an assistant professor of management at Northeastern University. He received a Ph.D. from Rensselaer Polytechnic Institute, a B.S. and M.S. in engineering from Northeastern University, an M.S. from Harvard University and an M.B.A. from Rensselaer Polytechnic Institute with a concentration in the management of technology. He has been associated with

Arthur D. Little Co., Cryogenic Technology, Inc., and Intermagnetic General Co., which specializes in superconducting materials and magnet systems. He served as vice president and general manager of Intermagnetic's magnet division.

John H. Vanston is president of Technology Futures, Inc., a management education and consulting firm that specializes in technology forecasting, strategic issues management, and the management of technical innovation. He was formerly deputy director of the Center for Energy Studies at the University of Texas at Austin and a member of the faculty of the mechanical engineering department. He is a graduate of the U.S. Military Academy and Columbia University. He earned his Ph.D. at the University of Texas at Austin.

About the Editor

Dr. Raymond W. Smilor is executive director and the Judson Neff Centennial Fellow at the IC² Institute, the University of Texas at Austin. He is also associate professor of management in the UT Graduate School of Business and executive vice chairman of the College on Innovation Management and Entrepreneurship of The Institute of Management Sciences.

He has published extensively with refereed articles appearing in journals such as *IEEE Transactions on Engineering Management, Research Management, Journal of Business Venturing,* and *Journal of Technology Transfer.* His research areas include science and technology transfer, entrepreneurship, economic development, marketing strategies for high-technology products, and creative and innovative management techniques. His works have been translated into Japanese, French, and Russian.

He is a consultant to business and government. He has lectured internationally in China, Japan, Canada, England, France, and Australia. He has also served as a research fellow for a National Science Foundation international exchange program on computers and management between the United States and the Soviet Union. He has been a leading participant in the planning and organization of many regional, national, and international conferences, symposia, and workshops.

He speaks extensively to business, professional, and academic groups in the United States. He is also involved in several civic and professional organizationns, and appears in *Who's Who in the South and Southwest.*

He is co-editor of six books:

Corporate Creativity: Robust Companies and the Entrepreneurial Spirit (Praeger 1984)

Improving U.S. Energy Security (Ballinger 1985)

Managing Take-Off in Fast Growth Companies (Praeger 1986)

The Art and Science of Entrepreneurship (Ballinger 1986)

Creating the Technopolis: Linking Technology Commercialization and Economic Development (Ballinger 1988)

Pacific Cooperation and Development (Praeger 1988)

He is co-author of two books:

Financing and Managing Fast-Growth Companies: The Venture Capital Process (Lexington 1985)

The New Business Incubator: Linking Talent, Technology, Capital, and Know-How (Lexington 1986)

Dr. Smilor teaches courses on technology and entrepreneurship at the UT College and Graduate School of Business. He is also one of the most highly rated instructors in the Management Development Program in the UT Graduate School of Business. He earned his Ph.D. in U.S. History at the University of Texas at Austin.